The Wolf Hybrid

Naima—alpha female at Wolf Park. Photo: Monty Sloan.

The **Wolf Hybrid**

Dorothy Prendergast

Second Edition
RUDELHAUS ENTERPRISES
Gallup, New Mexico
1989

Printed in the United States of America

Published by Rudelhaus Enterprises
Post Office Box 1423
Gallup, New Mexico 87305

ISBN 0-9623640-0-2

Front Cover: Rachel, reported 72% Tundra/Malamute cross owned by Heather Prendergast.

Back Cover: Adolph, reported 69% Tundra/Malamute cross. Photo: D. Prendergast.

DEDICATION

To Heather, the lady who has grown with our Wolf Hybrids and loves them all; who has had an active part in the publication of two books and the continuous production schedules of the magazine, The Wolf Hybrid Times; who lends her energy and enthusiasm to each effort; and remains a fine lady—my daughter.

ABOUT THE AUTHOR

Owner and breeder of Wolf Hybrids for more than thirteen years, the author and her family share their northwestern New Mexico home with a pack of Wolf Hybrids that has ranged in number from four to as many as twenty-one.

In its first edition, *The Wolf Hybrid* proved so popular that it was out of print within two years of its availability. Exemplifying her firm belief that no-one ever knows all there is to know about wolves or Wolf Hybrids, rather than simply having the first edition reprinted, Prendergast set about expanding and revising the book and expects that further revisions and expansions will be forthcoming in the future.

Prendergast has authored numerous articles which have appeared in periodicals and newspapers nationwide and, since 1986, has been the editor/publisher of the magazine *The Wolf Hybrid Times,* a magazine about wolves and Wolf Hybrids distributed internationally by Rudelhaus Enterprises.

She is also one of the founders of the Wildlife Education and Research Foundation, an organization which conducts beneficial research projects on the wolf and Wolf Hybrid.

Rudelhaus, as one would expect, loosely translated from German means a house or place inhabited by a pack of wolves, in this case, Wolf Hybrids.

TABLE OF CONTENTS

PREFACE

A low, mournful sound pierces the predawn silence, followed by a second, more melodious howl. Soon the air is filled with song as the sky slowly begins to lighten to a dark, grayish light.

It is the "hour of the wolf"[1] and the Wolf Hybrids are calling to each other and to their owner, signaling the ending of darkness and their impending feeding time.

Each voice is different; each voice distinguishable from the other. The chorus lasts for a few minutes and then subsides as the animals begin moving about their areas and remarking their territories, but the intermittent swell of their voices will continue until they are fed.

The sun is over the horizon. Their stomachs satisfied, the vocal communication subsides into physical interaction between the animals.

Is it the song, the beauty or the myth that makes the wolf so intriguing? In recent years, the fascination with *Canis lupus* has become more and more evident, with the publicity geared toward helping the recovery programs of the wolf in the wild, the many films and commercials about (or using) wolves and Wolf Hybrids, both semi-factual and fictional, and the phenomenal increase in production and sales of Wolf Hybrids as reflected in the want ads throughout the United States.

Zoos and breeders are besieged by calls from people trying to buy pure wolves, only to find that private ownership of these animals is not only not recommended, but in most cases, illegal. As a result, the Wolf Hybrid has rapidly become one of the most sought after animals in the United States.

* * * * * * *

Akili and Kaleah, two of the pure wolves at Wolf Park. Photo: Monty Sloan.

CHAPTER I. THE WOLF HYBRID MYSTIQUE

Wolf Hybrids, as a breed, are impossible to describe. To date, there has been no universally agreed upon standard for the breed, no general agreement of which breeds of *Canis familiaris* should or should not be used for cross-breeding, no agreement among breeders on goals for appearance, size, temperament, nor on what percentage of wolf blood such a "breed" should have. In short, there has been no movement strong enough to standardize breeding lines so that a true "breed" could be established. The only common demoninator among owners seems to be that they *think* they want an animal with the highest percentage of pure wolf lineage as is legally possible, regardless of its size, looks or personality. Compounding the problem are the endless numbers of "backyard breeders" who will cross-breed with just about anything, thinking that the progeny will still look "wolfy" but perhaps have a more mellow personality.

With such an awesome array of choices, not only breeds of dog but of subspecies of wolf, the results of such cross-breedings often bear very little resemblance to the animals whose name they bear. A Wolf Hybrid that is a Chow cross has very little in common with a wolf/Malamute or wolf/Husky cross. When through various generations, many different breeds of dogs are used, the resulting offspring often bear little resemblance to a wolf of any subspecies.

"Canis" is the scientific common family name for all domestic dogs (*Canis familiaris*), wolves (*Canis lupus*), coyotes (*Canis latrans*), dingo (*Canis dingo*) and jackal (*Canis aureus*). They are all part of the family *Canis*. Through natural breeding and environmental and hunting demands, various species developed.

In the species *Canis familiaris*, the direct relationship to the wolf is most obvious in the development of Huskies and Malamutes as domestic breeds. People are accustomed to seeing pictures of them pulling sleds full tilt in the dead of winter and most are familiar with the fact that the breeds developed from the cross-breeding of village working dogs and wild wolves in Alaska and the frozen arctic zones of Asia. The wolf was desirable because of its stamina; it was undesirable, according to the Eskimos, because it was inclined to work more reliably at night than during the day, and was easily distracted. The combination of wolf and Husky types, however, produced superior working animals. After many years and much work by a handful of devoted breeders and sled racers, the American Kennel Club finally recognized the breeds of Siberian Husky (1941), Samoyed (1957), Malamute (also listed as Alaskan Husky) (1960), Belgian Sheepdog and Turvuren (1959). According to Mario Miglionardi, German Shepherds were still classified in England as "Alsation Wolf Dogs" until as late as 1930.[2]

The Eastern Timberwolf (*Canis lupus lycaon*), is one of the smaller subspecies of wolf, though often touted as being one of the largest. It is lean with comparatively long, slender legs, is fairly shaggy looking and is perhaps the most well-known of the subspecies. Most of the "timberwolf" subspecies, also including the Northern Rocky Mountain Timberwolf (*C.l. irremotus*), Texas Gray Wolf (*C.l. monstrabilis*), southern Rocky Mountain Timber Wolf (*C.l. youngi*), etc., are now generally grouped into a common category, "the Gray Wolf," though there are some quite noticeable differences between them.

Canis lupus tundrarum, variously referred to as "tundras," "arctic wolves" or "Alaskan wolves," on the other hand, are "large, with long, usually light-colored pelage, though as with most subspecies, there are some dark phases. It is closely related to *C.l. pambasileus* but is paler and grayer."[3]

The Mexican Wolves (*C.l. baileyi*), the smallest of the subspecies, are beautiful animals with softer pelage and often coppery tones running through their fur. The rather large coyotes in the southwest are often mistaken for Mexican wolves, a fact which, because of the large number of coyotes in the area, compounds ranchers' already stiff resistence to reintroduction of wolves in the southwest. The huge buffalo wolf (*C.l. nubilus*), with its long, coarse and bushy coat, seems enormous by comparison.

These are but some of the subspecies listed by Hall and Kelson[4] in

1959. The subspecies were often determined and named, however, on the basis of as little as one specimen killed in a particular area and on skull measurements, with a projected area of territory being surmised rather then measured.

All the subspecies have one thing in common; the piercing lemon-colored eyes that seem to see right through you and that look almost translucent.

The intent, "staring eyes" are perhaps the greatest reason for the persistent myths, fear and/or worship of the wolf through the centuries, promoting fears which have nearly eradicated the wolf from the earth and which have all but prevented factual and rational study of the animal as it really is.

Wolf lovers glamorize the wolves—making them as nearly akin to a superbly social-conscious human as possible. People who fear them see them as vicious killers of man and beast—living transformations of the devil; something to be killed and forgotten. Neither version is correct and, after pouring through volumes of what so-called scientific data is available on wolves in the wild, one who has been around pure wolves for any extent of time can only shake his head in amazement at the paucity of really factual information that is available to the general public.

One of the better compilations of legends about wolves published to date is *Of Wolves and Men*[5] written by Barry Holstein Lopez. After giving a brief summary of published data on wolves in the wild, the author devotes the major portion of the book to legends and folk tales which have helped to build the aura of fear, love and respect of the wolf in modern times.

Until recently, few researchers have had continuous, physical contact with wolves in the wild on a day-to-day basis, making available factual literature scarce and often erroneous. The works of Dr. Eric Klinghammer, Patricia Ann Goodmann and the collected works edited by Harry Frank[6] on captive wolf research have probably shed more light on behavioral patterns of the wolf in the wild than any others because it is based on continuous observation throughout the animals' lives rather than limited and mainly long-distance opportunities for observation.

Given the general fear of wolves, the lack of factual, unromanticized information on the wolf side of the Hybrid, why do people want to own one?

People who want to own a Wolf Hybrid have many reasons, but far and above any stated reason, is the enchantment of owning, living with or controling a "part wolf"—i.e., part wild animal. The underlying statement in this obviously has very little to do with relating to the animal on a one-to-one basis, or what the animal looks or acts like. In some cases, the somewhat macho desire for power or control becomes translated into almost a social status symbol and fierce, jealous rivalries are rampant between breeders, organizations and clubs throughout the country.

There are those who are sincerely concerned about the disappearance of the wolf in the wild and feel that if it is illegal or impossible to preserve pure wolves under private ownership, then the next best thing would be to preserve as many of the wolf characteristics as possible in the Wolf Hybrid.

Others feel that the reintroduction of wolf blood into certain breeds of dog would be the best way of developing a superior breed in relation to stamina, health, looks and utilitarianism.

And there are those of us who, having lived with a Wolf Hybrid, could simply not settle for anything less.

Thoren—British Columbian wolf owned by Kent and Dana Weber. Photo: D. Prendergast.

CHAPTER II. WHAT BREEDS OF DOG SHOULD BE CONSIDERED?

The answer to this question is purely one of personal preference. The obvious mixes for obtaining a good, stable and outgoing Wolf Hybrid with a wolf-like appearance and conformation would be the domestic breeds more closely resembling the wolf—the Huskies, Malamutes, Shepherds and perhaps, Samoyeds. To cross to a short-haired or softer haired breed, a short legged or smaller breed simply makes no sense because of the added "undesirable" and less wolfy characteristics to be inherited.

A pure wolf typically has coarse, long hair with a very thick undercoat and a long, fairly straight tail that is carried straight or down rather than up, except in times of excitement. The softer, silkier hair of most domestic breeds and the curled or "sickle" tail are very hard to breed back out of a line. In many breeding lines utilizing Collie influence, the long hook nose is never lost.

Many people cross with German Shepherds, hoping for watchdog or guard dog qualities. Unfortunately, this often backfires for several reasons. A pure wolf, by nature, is fearful and will much sooner run from strange humans, given the opportunity (except, of course, where puppies are concerned). This fearful trait, combined with the characteristics of a highly bred German Shepherd often throws very shy, nervous and distrusting cubs who are more prone to running and hiding than defending property whenever anyone is near. On the other hand, I have seen some very reliable wolf-cross guard dogs of a very low percentage and which look and act like regular German Shepherds and have wondered about the reason for the introduction of or claim of wolf blood, given the purpose for which they are being sold.

Many breeders, rather than speaking in terms of percentages, prefer

Denali—reported 56% Timberwolf/German Shepherd. Photo: Darlene Sparks.

to emphasize "genetic blends" or "gene pools," wherein they are speaking of the various (sometimes many) combinations of dog breeds and wolf subspecies which have contributed to the producing of a particular litter. Where breeding has been carefully and intentionally controlled for the development of specific traits, some beautiful and somewhat more stable high percentage Wolf Hybrids have been produced using combinations of cross-breeding. The "genetic blend" reference, however, has also been used as a blind against providing accurate lineages.

The "ideal," as many breeders see it, myself included, is a wolf/dog cross strongly and attractively resembling its wolf counterpart visually, slightly aloof and territorial, but easily managed by its owner. It should have a coarse, heavy coat with the thick undercoat, a nearly straight tail with the scent gland marking well defined. The snout should be slender and straight, or arced slightly upward as in their wolf ancestors; the eyes, lemon to amber colored. The ears should be straight and semi-rounded at the ends. The body should not be "heavy" appearing. Rather, the legs should be long and lean in comparison and the body appearing lythe and agile. The hips should be lower than the shoulders, but this should not affect the overall smooth gate of the animal. At a trot, it should "track" at an angle.

Universally, the animals should never be "ugly," nor have an aggressive or "mean personality. If one or both of the parents exhibit either of these traits, it would not be wise to buy the cubs.

Following the adage that "more is better," many people seem to expect and demand larger animals, another throwback to "the *big,* bad wolf." A pure female wolf will rarely exceed 95 pounds, more commonly being 60—85 pounds. A male does not often exceed 110 pounds in any subspecies.[5] In Wolf Hybrids, the dog influence is what adds pounds and inches, and breeders should make buyers aware of and emphasize to prospective buyers that it is the *dog* and not the wolf influence that adds the volume to the animals.

The question is often asked why the American Kennel Club (AKC) does not recognize the Wolf Hybrid as a breed. The AKC has fairly rigid standards for the breeds it recognizes, the primary among them being able to reliably reproduce the "standard for the breed" (including color variation) in entire litters by any breeding pair. Obviously, a fair amount inbreeding is required to accomplish this—one thing that many wolf opponents criticize about the wolf in the wild. They argue that such inbreeding makes them fierce, high-strung or inferior, when actually, in a wolf pack in the wild, only the fittest of those produced survive.

The requirement for uniform reproduction of the standard is a primary reason why it is unlikely that the AKC will ever recognize the

Sasquach—reported 94% Alaskan Timberwolf/Malamute. Photo: Lynne Williams.

Wolf Hybrid as a breed, even if an enforced goal of looks, conformation, temperament, coloration, breed of dog and subspecies of wolf should be used, and relative percentage of wolf blood desirable in such a breed can be universally established among breeders.

While it seems it should be a simple thing to accomplish through generations of strict breeding, within a pure wolf litter, the cubs will exhibit as many colors and variations in markings as there are cubs, and this seems to be one trait that is not lost through cross-breeding, no matter how much the wolf blood is diluted. Since serious attempts at developing uniformity are a fairly recent phenomena, it is a rarity to find a breeder that can produce consistent litters of uniform color and markings, no matter how controlled the breeding may be. A single litter of Hybrids will most often have a range of colors from black through brindle, red, gray and white, and with a variety of masks and markings.

Likewise, it is impossible to predict what color a cub will be when mature, or if it will retain any of the markings it carries the first few months of its life, although a high degree of "wolfiness" can be fairly easily maintained. Even with carefully controlled breeding practices, there is the indeterminant heredity factor that breeders cannot control. It is not at all uncommon for pups to throw back not to the parents, but to the grandparents and great-grandparents (just as in humans) and you may find one out of every litter that bears as little resemblance to a wolf as to a Labrador and vice versa. You may come up with a very dog-like animal in appearance that cannot be treated any other way than as a pure wolf, and you may also come up with some very wolfy looking pups that have absolutely no "wolf" characteristics personality-wise. Size and temperament are more easily controlled by knowledgeable breeding practices than certain physical traits, although as will be discussed later, personality is shaped, to a large extent, by what happens after birth.

CHAPTER III. CONSIDERATIONS FOR THE PROSPECTIVE OWNER

In reality, there is no such thing as a "Wolf Hybrid." A "hybrid," by definition, is unable to produce replicas of itself and is in most cases, unable to reproduce at all. What we are actually talking about are wolf/dog crosses.

Animal lovers are all subject to the same emotional appeal—that cute, furry little puppy. "Isn't it CUTE!!" but few prospective owners really think much about what the animal is going to be like when it grows up. Or whether they are indeed going to keep the animal for life. I recently read an article pleading for sterilization of animals, citing the overflow and subsequent destruction of pets at dog pounds and humane societies throughout the country. It quite correctly referred to the very common attitude of "if it doesn't work, I can always give it away" as having "disposable pets."

Prospective owners should think quite seriously about whether they want a rather independent, decidedly different 75-100 pound animal barreling around the house; whether they can cope with the special behaviors of a Wolf Hybrid; whether they can provide proper containment facilities for it. Many people purchase a small puppy and do not have the ability to provide the space that such a large dog needs, let alone the special containment facilities needed for a Wolf Hybrid.

Most people are not prepared for the voluminous amount of hair their homes will acquire at least twice a year. Wolf Hybrids, like the wolf, have a very heavy, thick undercoat and it, along with the guard hair, is shed out at the end of winter and the end of summer, with the spring shedding being the heaviest.

Again, a point which must be emphasized again and again, many people purchase Wolf Hybrids with no forewarning at all of the special

9

behaviors or traits they may encounter as the animals mature, deciding suddenly one day that they cannot cope with the intelligence, curiosity, what they see as "destructiveness" or the adolescent behavior. As is rather comically evidenced from the following article published in *The Wolf Hybrid Times* (February 1986), one must be a rather unique individual to love and cope with such animals on a daily basis.

Cree—reported 77% Arctos/Tundrarum/Sherpherd/Husky. Photo: Bonnie Rashleigh.

PSYCHOLOGY OF A WOLF HYBRID OWNER
(A SENSE OF HUMOR IS REQUIRED)

If you're going to have a Wolf Hybrid, you'd better have a sense of humor. Whether you have a low percentage or high content Hybrid, it saves a great deal of frustration!

When you see your favorite sweater in shreds, when parts of your electronic equipment finally turn up when you're cleaning the yards, when you (and your neighbors) can't sleep at night because of the howling, when you wake up at dawn to a VERY COLD nose poking you in the face, you'd better have one.

Why do we have these animals who are determined to drive us crazy? We and thousands of others would like very much like to know why we put up with all this—*and like it!*

I was trying to build a shed. Nothing elaborate, but a storage area where tools and some small equipment could be kept in safety. Being the somewhat disorganized amateur do-it-yourselfer that I am, I was having a hard enough time keeping track of my tools and materials without any outside help.

I became aware of part of my problem when one of the Wolf Hybrids went barreling by me with a 10 pound sledge hammer dangling from her mouth. While chasing down the sledge hammer, I noticed other missing tools, nails and assorted materials stashed about—some partially buried. This experience filled me with firm resolve that my next project would be to build more pens so that I could work on my projects in peace.

Build more pens I did. One of the pens was to hang my laundry in so that we could enjoy the smell of air dried sheets and clothing that were all in one piece. There are pens around the trees. There are pens around other pens where the chickens are scavenging for grasshoppers; pens around gates to keep them from sneaking out when I'm trying to sneak in; and gates in the house to keep them from clearing possessions from especially vulnerable rooms. The quantity of pens now outnumbers the Wolf Hybrids, but I'm smiling!

I smile a lot too when I'm cleaning the yards and they're all trailing behind me leaving fresh deposits to replace the ones I've just cleaned up. I don't smile as much when after I've been entertaining a strange animal. They feel the absolute need to remark their territory, particularly if that territory happens to be in the house. You learn to develop a system of who's allowed to be in the house after who, or else spend your time like a scullery maid scrubbing the walls and floors after each visit.

You need a very special and different sense of humor to tolerate the intense jealousy and sometimes downright hatred of one animal for another. I often find myself feeling personally insulted by the demands to choose one over the other or of being knocked down in the frenzy of trying to be THE special, preferred animal.

At one time, I had two females who were half-sisters—same parents but different litters—who absolutely wanted to exterminate each other at any opportunity. The jealousy and hatred was so intense that they had to be kept separated by solid walls and fences, after several terrible and expensive battles—one through the chain link fence in which both were severely injured. The purely emotional injuries I suffered from these fights were sometimes as bad as the physical damage to the animals. And again I asked myself, *"Why* do I go through all this?"

Your sense of humor grows and is fed by the daily reinforced knowledge that no matter how long you have raised Wolf Hybrids or how much you think you know about them, you actually know almost nothing, and that which was true yesterday, probably isn't true today. You find a trustworthy person to care for them when you have to be gone a day or two, and spend much time making sure that the animals know, like and trust the person, only to have them turn tail and hide the next time he comes around.

After four years of living with the animal you confidently tell someone that "this dog never barks" and just to confound you, he punctuates the sentence with high pitched yapping, leaving you muttering that "it's not a regular type of bark."

When listening to so many people talk about what great mousers their Hybrids are, I try hard to *keep* from smiling when I think about why we borrow the neighbors' cat every so often to clear the house of the furry little creatures. Some of my Hybrids have taken great interest in watching the little fellows run around, squeaking in fear of the hulking shape with yellow eyes watching them so intently. But that's about as far as it goes—great interest.

On the other hand, my good mousers also tend to be even more skilled thieves, and cannot be left in the house alone or things start disappearing out the dog door in rapid succession. All of which results in borrowing Creaky (the cat) for a few hours every so often while the Hybrids are penned out.

J.D. Wolfe, percentage unreported. Photo: Corienne Cherry.

Oh yes, a sense of humor is very helpful. It ought to be made a mandatory requirement to own a Wolf Hybrid. You will find it helpful not only in living with your animals, but in restraining yourself when talking to other people about them. Tolerance soon gives way to exasperation when you have listened to "factual information" about wolves and Wolf Hybrids from people who have never even seen one. Just keep on smiling....

Almost without exception, most first time purchasers of Wolf Hybrids want the highest percentage animal possible or, if that is not possible, then at least assurance that the animal is half wolf, the assumption being that one of the parents is a pure wolf. Herein lies a double-edged problem that probably lands many Wolf Hybrids and their owners in deep trouble and creates a lot of misunderstanding,

13

both among the ranks of owners and the governmental agencies in charge of regulating animal ownership.

Led on by the glowing descriptions of some breeders and memories of movies depicting anthropomorphically endowed wolves, people buy high percentage or first generation crosses because "they want the wolf personality" when actually, they neither want, nor could in any way deal with a pure wolf's personality. What they are really envisioning owning is a very docile dog that looks like a wolf.

A pure wolf, while extremely adaptable, is still a wolf and subject to all the demands its body, environmental, instinctive and inherited behavioral traits place upon that body. Chemical changes take place within the wolf's body in different seasons of the year, particularly in mating season, and the changes in chemical balance affect behavior, just as they do in humans. Wolves are Canids in their purest form—predators, animals which must defend their territory (however limited it may be), animals whose behavior will be subject to change during mating season and whose behavioral patterns and needs must be respected and understood in relation to the animal and its inherited behavioral patterns themselves. They are not and could never accept being "pets."

It is for this reason that owners or breeders speak of "socialization" of wolves rather than "domestication." Pure wolves, particularly those bred and raised in captivity, may be accepting and trusting of humans and some of the requirements imposed on them by humans, so long as the humans are also responsive to their behavioral codes. But once the bond of trust has been broken by a human trying to supercede the dividing line between *Canis lupus* and *Homo sapien*, that bond is often impossible to reestablish, particularly since most humans tend to require that such a relationship be on *their* terms and are unable to relate to the behavioral characteristics and psychological responses of the animal itself.

High percentage Hybrids, F^1 and even F^2 generations are not exempt from the problem, and it is, therefore, wiser for the first time owner to acquire a lower percentage animal, further removed from the direct pure wolf cross, with a generally mellower personality than to jump into a situation he may discover as time goes on, is not at all what he wants in a companion nor is prepared to deal with in a manner that the animal can accept.

CHAPTER IV. CHOOSING A WOLF HYBRID

The potential buyer of a Wolf Hybrid should begin first with making an educated decision as to whether he or she really wants to have one. In particular, first time owners should spend a good deal of time observing, researching and discussing behavioral traits and growing patterns with a number of longtime owners of both high and low percentage animals. What one person may thoroughly enjoy in an animal may be absolutely intolerable to another. New owners should be made aware of things that may be problems for them, though others do not see them as such. (See Chapter III.)

The buyer should have a firm idea of what he or she wants in and from an animal and then seek breeders providing those desired qualities. I must emphasize the words "desired qualities" as opposed to "percentage" because often the two have little in common. With the fast-growing number of concerned and reliable breeders and concerned organizations, it is no longer necessary to take a chance on what one is buying or to buy an animal simply "because it is part wolf." (See Appendix A.)

MEET THE PUP'S PARENTS

In purchasing any animal, you should compare the parents of the pups with the ideal you have in mind and compare adult wolf/dog crosses of different dog breed lineage. Do you prefer the leaner but often more skittish German Shepherd crosses? The stockier, bushier Malamute and Huskie crosses? Is color a problem? (Adults are rarely the same in coloration or markings as when they were five week old pups). Are you concerned more about registration papers than how the resulting adult will look and act?

Insist on photos of the parents and, if possible, the grandparents and grown pups from previous litters. Try to meet in person some of these other progeny. Check on their personalities, conformation, color, texture of coat, and demand a guarantee in writing on your pup's bloodlines and against dysplasia and other inherited defects. Remember that the pups will be very much like their parents and if you are unhappy with or unsure of one or the other, don't buy the pup.

Demand proof and documentation of lineage and percentage, including each and every subspecies of wolf in the line and each and every breed and/or cross-breed of dog, the name of each and the name and address of the last known owner of that animal. Any responsible and knowledgeable breeder can and will provide you with this information. IF the animal is registered, the breeder should provide you *immediately* on purchase with copies of the parents' registrations (including proof of their lineage) and forms with which you can register your animal without your having to provide further information, except your name and address, to the registry.

Do not in any case buy a puppy whose lineage is undocumented or that you cannot compare the parents with the ideal you have in mind. I have seen many *supposed* "part-wolves" which show very little indication, if any, that there is a drop of wolf blood in them—only the wishful thinking of some very naive and perhaps macho owners. As Editor of *The Wolf Hybrid Times,* I receive many letters and calls from owners who belatedly find out that what they have bought is far less than what they thought they were purchasing. It is unfortunate that, along with the out-and-out unscrupulous breeders (who also exist), there are many, many unknowledgeable "backyard breeders" who firmly believe that they have very wolfy, high-percentage animals, who have no idea of how to calculate percentages, no idea of what pure wolves actually look like, and who sadly, sell their pups to unsuspecting, unknowing people as being "wolfy, high-percentage animals." Ironically, there are also some rather idiotic people who will sell very high percentage animals to unsuspecting buyers representing them as being lower percentage and more dog-like in personality in order to "get rid of them." These animals subsequently often end up being disposed of when the new owners are unable to deal with them.

Again, I can't suggest too strongly that you are certain of what you are buying and be able to document it before actually purchasing it.

You should also make certain that the puppy has been adequately immunized and wormed and, if it is five weeks of age or more, that it is already on a regular worming and immunization schedule against ALL the diseases to which it is most susceptible. (See Chapter XIX). This information should also be certified in writing, detailing the date and the schedule of immunization which should be followed after taking the puppy home.

CHAPTER V.
CALCULATING PERCENTAGES
(GENETICS)

Accurately stating the percentage of wolf blood in a Hybrid can be very confusing. This is particularly so if an unscrupulous or uneducated breeder tells you that this pup out of a given litter is only 50%, while that pup from the same litter is 98%, the breeder's reasoning being that the first *looks* more like a 50% to him, while the

Choctaw—reported 65% Timberwolf/Alaskan Malamute. Photo: Jon O. Youngstrom.

Figure 1

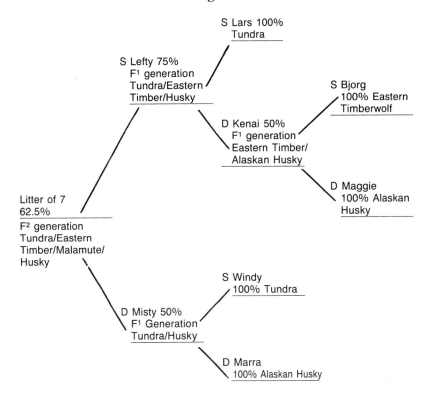

S Lars 100%
Tundra

S Lefty 75%
F¹ generation
Tundra/Eastern
Timber/Husky

S Bjorg
100% Eastern
Timberwolf

D Kenai 50%
F¹ generation
Eastern Timber/
Alaskan Husky

D Maggie
100% Alaskan
Husky

Litter of 7
62.5%
F² generation
Tundra/Eastern
Timber/Malamute/
Husky

S Windy
100% Tundra

D Misty 50%
F¹ Generation
Tundra/Husky

D Marra
100% Alaskan Husky

last really looks and acts like a pure wolf. Scientifically speaking, this is simply an impossibility. (See Appendix B.)

Many breeders pin the percentage of wolf blood to figures that seem quite bizarre, such as 89%, 78%, etc.; however, assuming that the percentages of the parents are correct, the method of arriving at those figures is usually quite simple.

The wolf percentages of the parents are added together and then divided by two. For example, if one parent is 3/4 wolf (75%) and the other is one-half wolf (50%), the resultant litter would be 62-1/8% wolf (which is often upgraded to 63% wolf). Figure 1 represents a typical charting of bloodlines and percentages. The breeder should provide you with such a pedigree and certify its correctness.

Returning to the original statement, however, those not educated in the field of genetics may become confused by breeders' claims. In the problem stated above, (that of claims of different percentages of pups in the same litter), the breeder is totally misleading the purchaser.

Shakaw—reported 81% Timber/Tundra/Siberian/Malamute. Photo: unknown.

This sort of judgment about percentage is based on "phenotype"— the visual and subjective presentation of the animal, rather than on its true lineage. I am often prone to make the comparison of a family in which there is a black South African mother and a blond, blue-eyed German father, and there are three children. Genetically, each offspring of this marriage has a 50-50 chance of inheriting the "phenotypical" characteristics of each parent. Each has the potential of phenotypically resembling a black South African, a blond, blue-eyed German, or a combination of the two (i.e., mulatto). But *each is still l/2 black South African and l/2 German.* The same genetic chances exist in each pup in any given litter of cubs.

For this reason, Appendix B, a series of articles which appeared in *The Wolf Hybrid Times,* is being reprinted in full in the hope that potential owners or potential breeders will not be misled in buying, claiming inaccurate percentages of or selling the animals which they have.

CHAPTER VI. LEGAL REQUIREMENTS

There is further reason for obtaining the guarantee and documentation of the pedigree referred to in Chapter V, other than merely ensuring that you do in fact purchase a Wolf Hybrid that is honestly represented.

In most states, it is illegal to own a pure wolf and, in some, illegal to possess a Wolf Hybrid over a certain percentage. Although it is highly unlikely, there may be an occasion when you would have to prove that your Hybrid is not a full wolf, or a Hybrid of more than a percentage allowed. There are, as unbelievable as it may seem, some states which ban ownership of the Wolf Hybrid altogether, and where there is no other avenue of banning them legally, seek to ban them under the auspices of the rather recent "Potentially Dangerous Animal Acts."

For this reason, the following listing of regulations compiled annually by The Wildlife Education and Research Foundation[7] is printed below.

Please note that this information was assembled in January, 1989, and may not be currently applicable. You should check into the current laws of the state in which you live, as well as the current local and county ordinances to make sure that ownership is permitted.

CURRENT STATE REGULATIONS
PERTAINING TO WOLVES AND WOLF HYBRIDS

Compiled January, 1989
by D. Prendergast

The Wildlife Education and Research Foundation

In January of 1988, a listing of state regulations pertaining to ownership of wolves and/or Wolf Hybrids was compiled and published in *The Wolf Hybrid Times*. The listing was drawn from responses received directly from the various state Fish and Game Departments, although several did not reply. Once again, we have written for current statutes and regulations, and the following are summaries of or quotations from the responses received.

Where an asterisk appears before the state name, the response received indicated that no change had taken place since their response of January, 1988; we have therefore printed the 1988 response verbatim. Two asterisks indicate that no response was received in 1989.

Please note that these regulations apply at the state Fish and Game level only. In some instances, local, municipal or county regulations may override the state regulations.

Please note particularly the language some states use regarding "potentially dangerous species." This leaves the illegality of possession open to the discretion and interpretation of the Game Commission when such cases are prosecuted.

ALABAMA:

"Alabama has no regulations or laws that would prohibit ownership or sale of Wolf Hybrids or pure wolves with the exception of red wolves." The 1989 response states: "The laws and regulations concerning ownership of wolves or wolf hybrids in Alabama have not changed since our last correspondence. They are not protected by Alabama Conservation Laws or Regulations and a state permit is not needed to possess a wolf or Wolf Hybrid in this state. At the present time no change is expected with regard to these animals."

ALASKA:

"Any and all wolves in Alaska are 'game animals' according to Alaska Statute. Temporary stewardship, but not ownership, of a 'game animal' may be achieved by obtaining an appropriate permit from this Department. By regulation of our regulatory board (Board of Game) we may not issue a permit to keep a 'game animal,' including wolves, as a pet.

Among the domestic species which may be kept as pets, imported, exported, sold., etc., without a permit are dogs (*Canis familiarus*). People possessing wolf/dog hybrids in Alaska are dealing with *Canis lupus x familiaris*. Our courts would decide whether or not such possession was legal."

ARIZONA:

"We have long considered that any animal resulting from the cross of a wolf (*Canis lupus*) and a domestic dog (*Canis familiaris*) is a domestic animal and is not subject to our jurisdiction. New live wildlife rules, now in final review by the Governor's Regulatory Review Council, make no changes whatsoever on whether something is or is not considered a hybrid....Further, we are not now proposing nor considering proposing any such changes." "(They)...may be under (the jurisdiction) of agencies such as county rabies-animal control departments." Pinal County has banned possession of Hybrids.

ARKANSAS:

The 1988 response stated: "It is unlawful to ship into or possess any species of wildlife in Arkansas which is designated undesirable by the Commission." "These animals (wolves) may be kept as pets if they are captured outside Arkansas and proper documentation is available." No mention of Hybrids. The 1989 response stated: "We have no plans to change any regulations pertaining to wolves in the near future."

CALIFORNIA:

"The ownership of pure wolves is illegal except by the few people qualifying for a valid permit from Fish and Game. Among the criteria for such a permit are rigid requirements for facilities and experience in raising such animals. F^1 generation Hybrids are considered as wolves. F^2 and subsequent generations of Hybrids are, and will be, considered domestic dogs and will not be regulated except in the same way as any other domestic dog."

COLORADO:

"Domesticated wildlife means those species...that would otherwise be considered wild native wildlife, but have been lawfully acquired and are held by private ownership and those which have been born of such lawfully acquired privately owned stock and raised in captivity....No person shall sell, acquire, purchase, trade or barter live domesticated wildlife unless he first obtains a proper license as provided in this chapter...Except as otherwise prohibited by Wildlife

Commission regulations a (Commercial Wildlife Park) licensee may buy, sell, trade, barter, propogate, import and export any lawfully acquired wildlife...." These regulations apply to pure wolves. There is no reference to Hybrids in the regulations except that F^1 cubs may not be sold without license.

CONNECTICUT:

The 1988 response indicates: "Connecticut laws governing wild and *potentially dangerous* animals prohibit the importation or possession of any pure wolf or Wolf Hybrid by a private individual." (Hybrid is considered a wolf). The 1989 response states: "No changes in either the laws or regulations prohibiting private ownership of wolves, wolf/hybrids and other animals considered potentially dangerous under Section 26-40a of the Connecticut General Statutes. No changes are foreseen in the near future."

DELAWARE:

"All wolves and other species listed as an endangered species are forbidden in the state of Delaware. There is nothing in the statutes about hybrids; therefore it is the assumption of the Enforcement staff that hybrids would be legal."

** FLORIDA:

"Pure wolves and Hybrids which are 25% or less dog require a $100 permit fee per year unless exempted as a researcher or breeder. Provisions include: owner is 16 years or older, has knowledge of the species requirements and is able to provide a proper diet, health care and exercise, caging meets minimum specifications and that the neighborhood setting is 'appropriate.'" There are yard containment requirements. Minimum requirements are less than those recommended in WHT.

GEORGIA:

"Unlawful to import, transfer, sell or possess (wolves) without first obtaining a wild animal permit...and shall be 'issued only to persons engaged in the wholesale or retail wild animal business or persons exhibiting wild animals to the public' (or scientific or educational purposes)." Liability insurance of $40,000 per animal up to a maximum of $500,000 is required. The 1985 response also stated: "A direct F^1 cross is considered the same as a wolf and cannot be kept.

Anything less than 50% wolf is considered domestic and requires no permit." The 1988 response also noted that "Wolves still cannot be possessed in Georgia as pets (or for other such personal reasons)." The 1989 response makes no reference to Wolf Hybrids and no specific reference to hybrids of *any* generation can be found in the regulations. One pertinent development, though not related to a law change, is the apparent unavailability of the required liability insurance for wolves or other "inherently dangerous" wild animals in Georgia, which may even prevent the possession, etc. of such inherently dangerous animals in this state for any purpose.

IDAHO:

"Any canine exhibiting primary wolf characteristics shall be classified as a wolf for the purpose of identification. All such canines shall be tattooed, registered and licensed by the Department of Fish and Game....Since wolf/dog hybridizations are known to exist within Idaho and hybrids are not protected by the United States Endangered Species Act, a biological evaluation shall be required of the animal to determine species priority before the Department of Fish and Game may take any action in accordance with the United States Endangered Species Act."

This evaluation must be done by a trained officer from Fish and Game, trained in the physical appearance of the wolf. Seven out of eight of the following characteristics have to be met for species priority:

1. Eyes shine greenish orange.
2. Ears rounded and smaller in proportion to those of the coyote.
3. Snout is broad with nose pad wider than one inch.
4. Legs are long, an adult would stand at approximately 26"-32" at the shoulder.
5. Length is 4.5 feet to 6 feet from the tip of the nose to the tip of the tail.
6. An adult weighs at least 80 pounds.
7. Tail is carried high or straight out when running.
8. Fur is long and coarse, varies from white to black but is generally grayish in coloration resembling the coyote. The underparts are not as white and the legs and feet are not as red as those of the coyote."

ILLINOIS:

"Please refer to 'the Dangerous Animals Act' for the State of Illinois. You will note the wolf is included and designated as a dangerous animal by the State of Illinois. There is no separate designation for the crossbred dog and wolf, and they are considered wolves in this State and must comply with this particular statute." The statute reads: "No person shall have a right of property in, keep, harbor, care for, act as custodian of or maintain in his possession any dangerous animal except at a properly maintained zoological park, federally licensed exhibit, circus, scientific or educational institution, research laboratory, veterinary hospital or animal refuge in an escape-proof enclosure." The letter goes on to say, "We do share your concern so that a citizen of the State of Illinois does not purchase or obtain such an animal and discover that they cannot possess the animal in a qualified complying manner. This type situation causes them embarrassment, legal confrontation and undue stress of the animal involved."

INDIANA:

"The regulations and statutes for Hybrid Wolf owners/breeders are the same as in 1988." In the State of Indiana, one who owns a wolf 75% pure or more must have a wild animal permit, which costs only $10. If an owner wishes to breed the wolves he/she must acquire a game breeders permit, which costs $15.

Each individual wolf must have a wild animal permit. This permit never expires unless ownership has changed. So, the owner who wishes to breed, in addition to his/her wild animal permits must receive the game breeders permit. The 100% wolf may be imported to the State of Indiana, but may not be released. Each application must be approved by a local conservation officer before receiving a permit."

IOWA:

(Response received 1/88) Possession of a wolf from a lawful source outside the state is permissible; however, "persons who own wolves which are considered indigenous to Iowa must have a Game Breeder's permit issued by the Iowa Department of Natural Resources if they own two or more." (Copies of regulations received in 1987 refer primarily to animal welfare requirements, such as housing, care, etc. Anyone who breeds and sells wolf x dog hybrids, which have any portion of dog, must be licensed and regulated under Chapter 162, Code of Iowa, and Iowa Departmental Rules Chapter 20. Our state, at city government level, is passing more local ordinances which prohibit the keeping of wolves and other so-called "exotics."

Response received 2/89) "No changes have been made to alter the regulation of wolves or Wolf Hybrids in Iowa. However, some cities are passing exotic animal ordinances and including wolves.

Under Department of Natural Resources, owners of two or more purebred wolves must have a game breeder's permit. Their facilities are inspected by personnel of that Department.

Under our Iowa Department of Agriculture & Land Stewardship, owners of wolf/dog hybrids (any portion of dog with wolf) must be licensed, if they sell, lease or exchange animals for a consideration. They must have four wolf/dog hybrids capable of breeding—any combination of males and females or females bred to some other owner's stud. These kennel facilities must be inspected and approved by our Department."

KANSAS:
"Kansas does NOT have any restrictions on wolves." (Hybrids were not mentioned).

KENTUCKY:
"(Wolves) cannot be imported or possessed except for certain educational, scientific, exhibition or research purposes approved by the commissioner." No mention of Hybrids. The 1988 response stated: "We have no regulations concerning Wolf Hybrids." However no mention of hybrids is made in the 1989 response.

** LOUISIANA:
(Editor's note): We have never received a response from this state except in 1988 when the note read: "We have no further information to provide at this time on the hybrid issue." However, we did receive word that there is an attempt underway to enact legislation similar to that of the state of Georgia, requiring incredible amounts of insurance for each animal, making compliance for legal ownership nearly impossible.

** MAINE:
The Commissioner of Inland Fisheries and Wildlife can issue a permit permitting one to "breed, rear or possess any wild bird or animal." ('any species of the animal kingdom, except fish, which is

wild by nature, whether or not bred or reared in captivity...') There is no reference to wolves or Hybrids; there are no provisions for fencing, feeding, etc.

MARYLAND:

"Essentially, unless you currently have a wolf or wolf hybrid in Maryland, it is illegal to acquire or own one now." The copy of the "Rabies Emergency Regulation" enclosed, reads: "A person may not import into Maryland any live raccoons, skunks, foxes, wolves, coyotes, bobcats or any other mammalian wildlife species or hybrids, for which there is no U.S.D.A. certified vaccine against rabies."

MASSACHUSSETTS:

"No wildlife (including wolves) may be kept as pets. They may be owned only for educational or scientific purposes or if the owners' primary livelihood depended on raising them (as in a fur farm). This would obviously not apply to wolves. Hybrids are not considered wolves, no matter what percentage, so long as the cross is with a domestic dog."

MICHIGAN:

"It is illegal to import or possess live wolves except for certain entities such as educational & research institutions, zoological gardens and others...Any 'wolf' having any number of genes from a domestic dog is considered a domestic dog and may be kept without a permit."

MINNESOTA:

"Minnesota does not currently regulate Wolf Hybrids, although they may be prohibited by local ordinances in some areas. However, if a Wolf Hybrid bears close resemblance to a 'pure' wolf, the owner may be required to show proof that the animal was obtained legally and that it is a hybrid. Pure wolves may not be taken from the wild, may only be obtained from properly licensed breeders and require a permit for their possession."

** MISSISSIPPI

No response has ever been received from this state.

MISSOURI:

"No person may keep a ...wolf... in any place other than a properly maintained zoological park, circus, scientific or education institution, research laboratory, veterinary hospital or animal refuge, unless such person has registered such animals with the local law enforcement agency in the county in which the animal is kept.'' ''A wildlife Breeder Permit is required to possess a wolf and/or wolf hybrids in Missouri.'' In addition, you must confine the animals in humane and sanitary facilities that meet standards established by the Missouri Dept. of Conservation and as specified in applicable federal regulations.

MONTANA:

(1988 response) ''There are no Montana statutes or regulations at this time pertaining to private possession of legally obtained wolves. Wolves held in captivity must be permanently marked with individual identifying tattoos. The Dept. of Livestock requires health certificates and importation permits for any animals brought into Montana.''

(1989 response) ''To date, no legislation has surfaced that would affect the status of wolves or Wolf Hybrids in Montana. Wolves and Hybrids, 50% or greater wolf, held in captivity are still required to be tattooed and registered with this office.''

NEBRASKA:

''A 1986 change in Nebraska revised statute 37-713 now makes it illegal for anyone to keep any wolf or crosses in captivity in this state.'' (The 1989 response indicates no change.)

NEVADA:

''The only regulation we have concerning wolves is that which deals with importation or exportation requirements. If the wolf is obtained from a commercial source or reared domestically, the importation or exportation requirement is waived. City or county governmental entities may have specific regulations or ordinances regarding wolves.''

NEW HAMPSHIRE:

''Wolves are not allowed to be imported...the offspring of a wolf-dog cross may be imported into this state as these offspring are no longer classified as wolves but as mongrel dogs.''

NEW JERSEY:

"Potentially dangerous wildlife regulations (which include pure wolves) precludes keeping them as a pet. They may only be kept for scientific study or exhibition. Wolf Hybrids are not regulated in any manner."

NEW MEXICO:

The 1988 response stated: "Ownership of pure wolves is illegal." No mention of Wolf Hybrids. The response received in 1989 states: "The situation in New Mexico is a little complex and certain aspects of the situation may seem conflicting...The reason for this situation is the controversy over the reintroduction possibility of an endangered wolf (*Canis lupus baileyi*) within New Mexico.

This situation is to some degree responsible for the following wolf possession policies:

Possession of Mexican wolf (*Canis lupus baileyi*) is prohibited. Possession of other wolves or Wolf Hybrids permitted. Importation of all wolves and Wolf Hybrids into the state for individual possession is prohibited."

NEW YORK:

"No person may possess, release, transport, import or export, or cause to be released, transported, imported or exported, except under permit from the department....any animal, the overall appearance of which makes it difficult or impossible to distinguish it from a wolf (*Canis lupus*) or coyote (*Canis latrans*)....Such permits may be issued only for scientific, educational or exhibitory purposes." ...No person shall, except under license or permit first obtained from the department, possess, transport or cause to be transported, imported or exported any live wolf, wolf/dog, coyote, coy dog....where the department finds that possession or tranportation, importation or exportion of such species of wildlife would present a danger to the health or welfare of the people of the state, an individual resident or indigenous fish or wildlife population. Conservation officers, forest rangers and members of the state police may seize every such animal possessed without such license or permit. No action for damages shall lie for such seizure and disposition of seized animals shall be at the discretion of the department.... The department may issue a license revocable at its pleasure to collect or possess...or sell... fish, wildlife ...for propagation, banding, scientific or exhibition purposes.

NORTH CAROLINA:

"Wolves are not specifically addressed with separate laws but would come under a "captivity license". North Carolina does not recognize or exert authority over wolves or Wolf Hybrids because of the inability to determine content and do not regulate them consequently."

NORTH DAKOTA:

Wolves, *Canis lupus,* are protected by state law. The season for taking wolves from the wild in North Dakota shall remain closed until declared open by proclamation.

Persons wishing to possess or propogate wolves in North Dakota must first obtain a permit from the North Dakota Game and Fish Commissioner and comply with laws as stated." No mention of Wolf Hybrids.

OHIO:

"The Ohio Division of Wildlife has no authority concerning wolves."

OKLAHOMA:

"The Oklahoma Department of Wildlife Conservations considers any wolf/dog hybrid as a domestic animal and *not* wildlife. Anyone can own an individual wild animal without having to purchase a non-commercial breeder's license. These animals have to be acquired from non-wild stock (i.e. commercial breeders) and a sales receipt must be kept to prove origin. If two or more animals are owned, then the non-commercial breeder's license would have to be purchased for $5.00. If two or more animals are owned and bred for resale, then a commercial breeder's license would have to be purchased for $48.00.

OREGON:

The 1988 response indicated: "There are no regulations for Wolf Hybrids." The 1989 response additionally provided procedures for applying for and requirements for obtaining a permit for the holding of pure wolves. Such regulations are obviously designed to provide secure, humane and educated housing and handling of any such animals.

PENNSYLVANIA:

The Game and Wildlife Code, Title 34 PA C.S., Section 2961 reads: "Definitions: 'Exotic Wildlife.' The phrase includes...wolves and any crossbreed of these animals which have similar characteristics in appearance or features. The definition is applicable whether or not the birds or animals were bred or reared in captivity or imported from another state or nation." The definition is modified, however, in Section 147.2 as follows: "Wolf Hybrid—any wolf hybrid not registered or licensed by the Department of Agriculture." There is a further handwritten notation on the response which reads: "A hybrid licensed as a dog by the Department of Agriculture is not exotic wildlife for purposes of our law."

RHODE ISLAND:

Wolves or Wolf Hybrids are not allowed to be sold or kept in the State of Rhode Island under any circumstances. The only exception being zoos, or botanical gardens which are licensed with the Federal Government.

* SOUTH CAROLINA:

"Any wildlife imported into the state must have a permit. No carnivores can be sold as pets. Such carnivores shall include animals known to be reservoirs of rabies, such as ...wolves." No mention made of Wolf Hybrids.

SOUTH DAKOTA:

Wolves and Wolf Hybrids less than three generations removed from the *wild* are considered protected animals and may not be possessed. Ownership of "domesticated" wolves permitted provided proof of ownership and purchase is provided. Hybrids not regulated.

TENNESSEE:

"No person shall purchase or hold live wildlife in captivity without first obtaining the appropriate permit as provided in this part. The annual permits and fees for holding live wildlife are....Personal possession—$10/animal or $100/facility. Commercial propagator (where more than $1,000 per permit year is obtained from the sale, barter, trade, etc. of the animals) ...Wolves (all species except domestic dog hybrids that are less than 75% wolf)...without having documentary evidence showing the name and address of the supplier of such wildlife and date of acquisition...The executive director shall issue a permit upon a satisfactory showing of qualification to possess live wildlife under the following conditions: 1) The applicant must be at least 21 years of age; 2) must have at least 2 years of experience in the handling or care of the Class I species for which the applicant is applying, or in the alternative, must take a written examination, developed and administered by the Agency evidencing basic knowledge of the habits and requirements, in regard to proper diet, health care, exercise needs and housing of the species; 3) ...Facilities for Class I animals may not be on premises of less than one acre and may not be located in a multi-unit dwelling or trailer park; 4) applicant must have a plan for the quick and safe recapture of the wildlife or if recapture is impossible, for the destruction of any animal held under the permit..." Additionally, "The permittee shall control and maintain Class I wildlife at all times in such a manner as to prevent direct exposure or contact between the animal(s) and the public..." Additional containment requirements are specified which include covered cages of ll-1/2 gauge steel chain link, or may be uncovered if fencing is at least 9 feet high with the top 3' of fencing inturned at a 45° angle.

"Owners of unpermitted wildlife who do not qualify for a permit to possess such wildlife shall dispose of such wildlife to an approved recipient within 30 days of notification by the Agency."

TEXAS:

"...Section 63.103, the only statute in the Parks and Wildlife Code pertaining to wolves (reads) No person may possess, transport, receive or release a live wolf in this state. Wolf Hybrids, coyotes and dog and wolf cross-breeds are not covered in this statute nor in any Department regulation. A person who has been issued an Endangered Species Propogation Permit by this agency may possess legally obtained purebred red wolves (*Canis rufus*) or gray wolves (*Canis lupus*) for the purpose of propogation and sale. The original propogation permit fee is $300.00 for the first year and $550.00 for a renewal that is good for three years."

1989 response indicates no changes to this law and none proposed.

** UTAH:

"For the time being, Utah does not give permits for the import or possession of Wolf Hybrids for personal pets. There are a couple of instances in Utah where there have been permits issued for use of wolf hybrids in movie making. Most all cities and counties in Utah will not allow Wolf Hybrids."

VIRGINIA:
"It is still illegal to import into Virginia any wolf. Cross breeds of dogs and wolves are not classified as purebred wolf and are not restricted by this agency. They would fall under the control of county animal control units."

VERMONT:
Wolves, coyotes and Hybrids permitted only for bonafide educational or scientific purposes. Importation and possession of wolves is illegal.

** WASHINGTON:
"It is unlawful to possess a wolf except by permit and permits are no longer issued."

WEST VIRGINIA:
(1988 response) "The wolf is considered a wild animal species native to this state and under West Virginia law, an importation permit would be required. An appropriate permit for whatever purpose it would be maintained after it was brought into the state would also be necessary. West Virginia has had some unpleasant experiences with such animals escaping or being released to the wild...Therefore, we will not issue an importation permit on a wolf." No mention of Hybrids.

(1989 response) "We will not issue a permit for a wolf."

WISCONSIN:
"Pure wolves are illegal except by permit for zoological, educational or scientific purposes or propagation for preservation purposes. Hybrids are not regulated."

** WYOMING:
Anything over 50% illegal; possession considered a felony. "1st progeny resulting from cross-breeding between an exotic or wildlife species with another species is considered to be true exotic or true wildlife."

CHAPTER VII. THE FERTILITY CYCLE

Contrary to popular belief, wolves are capable (given favorable environmental circumstances) of producing litters within the first year of their lives, especially where the wolves have been bred and raised in captivity. Also, contrary to popular belief, wolf cubs are not *always* born within a given number of weeks in the year. Many environmental factors enter into creating the fertility season, both in the wild and in captivity. Given differing latitudes, elevations, climates and availability of food and mates, and, in captive situations, the artificial altering of natural conditions through electrical lighting and heat, it is not unheard of for pure wolves to produce in what is generally considered "off-season."

Likewise, Wolf Hybrids, no matter what their percentage, may be "interested" and fertile at ten months of age. They may also come into heat more than once a year.

It is unwise and undesirable to breed wolves, Wolf Hybrids, or dogs during their first cycle of reproductive capability. The females may physically demonstrate a "heat"; the male may be more than willing to mount her, and if these are not symptoms of "false heat," "false fertility" or simply a "dominance exhibition," they produce a couple of pups. It is likely, however, (particularly in wolves or high percentage Hybrids) that the reproductive organs of the female are not fully developed at this age or that 1) the female is exhibiting a "false heat" and 2) that the male is simply extending the natural development process of "doing what he is supposed to do," but is infertile.

Some pregnancies do occur, but sadly, the pups are often underdeveloped, very few in number, or die shortly after birth. There may be great difficulty in the whelping process as well as fear of it, and subsequent rejection of the pups. Worse, because the reproductive

Mom and her cubs, percentage unreported. Photo: Linda Cary.

organs are not developed fully, they may be damaged to the extent of inability of the bitch to produce live pups again.

CONSIDERATIONS BEFORE BREEDING

In selecting breeding partners, one must be even more choosy than when selecting a puppy. The conscientious breeder is seeking to develop the best characteristics possible in the resultant litter. Ideally, the breeder should be very familiar with both the proposed sire and dam and have tracked down other siblings from the same litters as the prospective parents, and hopefully some of their ancestors in order to determine what he might expect from the litters to be produced.

He should seek qualities such as actual "wolfy" looks, conformation and gait, but with a more even, accepting (yes, "domestic") personality. My feelings about this are rather strong because most first-time owners could never anticipate or cope with the behaviors of a pure wolf nor a high percentage or first or second generation cross.

In any breeding pair, it should be determined by X-ray that neither they, their parents nor grandparents are predisposed to hip dysplasia or any weakness in the back or legs. All should also have a clear record of any other negative genetic predispositions.

A knowledge of the grandparents, and if at all possible, other progeny from other litters of the parents, is necessary to be able to evaluate the choice of mates. One cannot expect an instant correction of a problem with a mate having the opposite extreme of a particular fault. It is more likely that some of the litter will exhibit the one extreme and some the other. (See Appendix B.)

Before you breed, consider the following suggestions.

1. The bitch should be of above-average stock and should have the best traits in markings, conformation and personality.

2. The male should also be above average stock and of compatible size and lineage (i.e., if you are using Husky as the dog cross in one parent, the other should also be a Husky cross).

3. Both should be free of predisposition to hereditary defects and this should be documentable.

4. You should be aiming to produce puppies that are better than either parent.

5. The bitch should finish being wormed about one month *before* coming into heat and should be current on *all* innoculations. The puppies acquire a natural temporary immunity from their mother only if their mother *has* that immunity.

6. The bitch should be at least eighteen months old before being bred for the first time.

7. If two unfamiliar mates are to be bred, the bitch should be taken to the male for the actual breeding rather than vice versa.

The breeding season itself can be quite alarming to inexperienced owners. When a female comes into heat, her personality changes. As in a human female, preceding estrous the chemical balance inside her body has changed radically, often causing behavioral problems. Her interest is focused on her own needs and desires. She may become very temperamental and aggressive toward other females. The battles may be quite fierce, and for this reason, the bitch should be separated from any other animals but the male she is to be bred to as soon as she is due to come into heat. Additionally, for obvious reasons, she should remain separated from contact with any other male than the one she is bred with until the heat is well over. Though she may be capable of and allow a tie, she may not be fertile at the time you think she is. It is not also unheard of for a bitch to be impregnated by more than one male, producing a "Heinz 57" litter.

The males also experience a chemical and personality change. Previously docile and compatible males may become quite aggressive in asserting their dominance and if the animals are kept in a "pack" situation, increasingly serious aggression may ensue. Challenges may be minimal for the rest of the year and involve little more than that—a slight challenge. During breeding season, however, it is more likely that such clashes will be of a serious nature, causing very serious injuries to all involved. It is, therefore, wiser to separate the animals during this time—better to be "sheepishly safe" than later sorry.

If the pair to be mated cannot be kept together for the entire breeding period, then the female should be taken to the male, rather than vice versa. She is less likely to reject him or be too nervous to conceive.

If the female is being introduced to the male for the first time and for a limited period, you will want to keep a watchful eye out, without interfering. Quite often, breeders are frustrated because a female will not accept a particular male and may work on tearing him to shreds instead. This is not uncommon and I have met quite a few "virgin wolves" who have refused any attempted breeding whatever, but who in the "off season," are very compatible with the males which have attempted to breed them.

If the bitch will not accept the male you have selected, another male should be considered, provided that the owner is certain a tie has not occurred with the first male and that the new male meets the same high standards. If the female has been with the male for some time, there is not too much cause for worry and the animals should be allowed to breed as often as they will and without constant supervision.

HOW CAN YOU DETERMINE WHEN A BITCH IS COMING INTO HEAT? WHEN IS THE BITCH FERTILE?

After the bitch has had her first heat and has established a heat cycle, it is most often simply a matter of watching the calendar. For the first couple of heats, however, many new owners find themselves suddenly surprised.

Aside from the visible manifestation of estrous (i.e., bloody vaginal discharge), by close observation an owner can be aware of an impending heat by observing behavioral and physiological changes in the bitch.

Competition between the females may begin to erupt if they are in a common enclosure. Like a "flirtatious young thing," the female about to come into heat will begin "courting" the male of her choice, carrying her tail high in the air; leaping and "dancing" with him and alternately turning and exposing her as yet unripened vulva to him. This behavior generally begins prior to any noticeable bloody discharge, but the odor of the clear to yellowish discharge which gradually begins to increase is quite noticeable to the males. There may be half-hearted attempts at mounting during this period.

As the vulva begins to swell and the discharge becomes bloody, the attempted mountings will increase. Generally, ten days after the first noticeable bloody discharge, is the optimum time for conception. The bitch may conceive, however, more than ten days following this date and she should be kept segregated from other males for as long as the discharge continues. The pair should be kept together for this time, preferably for the duration of the heat cycle to be sure that conception has occurred.

Assuming that the breeder *knows* when conception occurs, the litter may be expected 61-63 days later.

Cub Scout—percentage unreported. Photo: Linda Carey.

CHAPTER VIII. PREGNANCY

Like the wolf, the normal gestation period in a Hybrid is 63 days, although individual females may vary a day or two. Four to five weeks after breeding has occurred, the female will be showing signs of pregnancy, with the swelling of the abdomen and general weight gathering all over the body. The nipples will gradually become more prominent, and finally milk sacks develop which make the nipples "hang." The pups in their sacks will also be palpable by an experienced breeder or veterinarian.

Early pregnancy becomes (behavior-wise) more uneventful, although the animals may become more affectionate and attention seeking, both to their humans and their "pack." As whelping time becomes more imminent, however, different behavioral patterns may be noticed. The bitch may begin digging a den. Some animals dig full-fledged dens more than six feet back into the ground, while others who accept and adopt the facilities provided for them may not dig at all, or only make a shallow depression in the ground big enough for the litter with enough air space to allow her to cover them without suffocating them. Some dig dens but at the last minute abandon them and have the cubs at another location half so protected. Some will strip their man-made "whelping den" of every foreign object it can, including artificial heating devices.

While the "instinctive cleaning" of the area may seem "cute" or just a vestigial response on the part of the bitch, actually all the efforts of humans to provide heating, nesting materials or light may be somewhat dangerous to the animals' lives. Many pups are unwittingly suffocated by rags, straw or scraps of newspaper put down by well meaning owners for bedding or litter material. The bitches most often take it out of the litter area, if possible, until the pups' eyes are open and they are moving about on their own, and often longer. I have seen them tear down electric lights, heaters and the pipes leading to them from high in the air in structures provided for whelping, which could have caused electric shock or fire. One of my bitches went so far as to pull out the pipes leading to the propane heater, all of which were supposedly out of her reach. She did this the night she had her litter, thus effectively eliminating any artificial heat in the whelping shed in what turned out to be -15° weather for over a week. Curiously, that litter of eight proved to be the healthiest and strongest of any we'd ever had and, while we worry about them every time, we have not attempted to provide heat in the whelping sheds since.

I have also seen bitches totally reject any artificial "den" provided for them and when prevented from using another area of their choice, "hold" their pups long after what would be considered a "safe" period, somehow refusing to whelp in a place they didn't want to... until finally their bodies could not hold them in any more. You may say "baloney—that's not possible." Veterinarians will tell you otherwise.

SPECIAL CARE OF THE BITCH

At the second to third week of pregnancy, the bitch's rations should be increased and split into two portions—fed half in the morning and half at night. A calcium supplement should be added to her diet, following the manufacturer's directions for her normal weight. This

Maicoh (reported 85%) and her litter. Photo: D. Prendergast.

will not only aid in the development of strong, healthy puppies and of the mother's milk supply, but will prevent esclampsia, a very severe condition, from developing after whelping. Vitamin supplements should be considered, especially if the normal ration consists of low protein dry food only.

Unless she is exhibiting some special problems, no other special care should be necessary for a healthy bitch. She needs exercise, she needs love and affection, shelter, a good diet and an ample supply of fresh water. In the last week or so, of course, like any pregnant being, she is apt to become slower moving and somewhat preoccupied by her abdomen, and should not be put under physical or emotional stress. She should especially not be made subject to stress within a pack. Rather, she should be free to prepare the whelping area.

WHELPING

If a breeder has been keeping careful records of the female, he will know what her normal body temperature is, which should be between 101.4° and 102°. If she is due and is becoming restless, panting, perhaps moving things from one place to another and her temperature has dropped one to two degrees below normal, she will soon begin whelping. Sudden loss of appetite is another sign of impending birth. Some bitches begin pulling fur from their chests or abdomens.

When a bitch is due, she may exhibit a wide range of behavior, depending on how close and dependent she is upon her human "alphas," whether it is her first litter, or how easily she can con her owner. A bitch may be super-independent and refuse any sort of companionship at all, whether with her mate or humans. When her cubs start coming, she may not allow either her mate or her owner to be near the whelping area. The same bitch, in a subsequent whelping, may almost demand attention and care during the pregnancy and whelping process. And again, some behave reliably during each and every whelping.

Just before the actual whelping begins, the bitch will lie down, whether in the den, shelter (or if she chooses, on your living room rug) and within an hour, the first pup should be born. The next two or three may follow within an hour or so of each other, but the last pups to emerge may take as long as ten to twelve hours to be born. Each pup is enclosed in a wet, semi-transparent sack and, as it emerges, the bitch will tear and eat the sack (generally severing the umbilical cord in the process). Do not attempt to take this away from her. It is very beneficial to both her and the pup physically. She will then lick the pup vigorously, which both dries it somewhat and stimulates its internal organs to operate in an oxygen atmosphere. She will then push and tug it to her nipples, whereupon it begins receiving the colostrum so necessary to its health.

Should the bitch not seem to know what to do, it may be necessary for the breeder to break the fetal sack, tie the umbilical cord with a clean piece of stout string or surgical thread and then cut the cord just beyond where you have tied it. If she still does not begin to lick the pup, you should dry it with a clean, soft, lint-free towel—gently if it is breathing; vigorously if it is not. When it is breathing freely, immediately put it to the mother for nursing. If she will not allow it to nurse, and if there is no nursing surrogate mother available, it is important that the owner be prepared with bottles, nipples and formula.

I will not attempt to give a presentation of "helpful suggestions" in case there are problems with whelping. There are many books filled with them. However, if a bitch seems to be straining, or if noticeable labor persists without the appearance of pups for more than one hour, call your vet. Unless you are trained for potential emergencies, you could do more harm than good, for both the pups and the bitch. The main thing to remember is not to panic.

At this point, I must inject a word of warning to the new breeder. Just as in mating season, you may find yourself facing a new and radically different personality in the bitch. While she wants and needs your affection and concern, she may very well drive you away from the whelping area after the pups are born until she begins to feel comfortable about moving away from them. This is especially true during her first whelping. While under other than breeding or whelping circumstances your bitch might very well die for your affection, you are, for awhile, going to have to stand second in line to her pups. Her fear of harm to them may extend to you too. This shouldn't be surprising—consider a human mother's reaction to any real or imagined threat to her children.

On the other hand, I have had some litters where the bitch really *needed* reinforcement throughout the whelping and rearing period. Some, through successive litters, simply had no idea what they were to do, were confused and somewhat alarmed by the mass of squirming, begging fur, or simply wanted someone else to get them through the whole process. One bitch literally had to be forced into caring for and nursing her newborn pups.

Give your bitch all the time and space she needs to take care of those instinctive reactions. If the whelping process has gone smoothly, she will generally let you know if something is wrong with those pups, given the opportunity. Nine times out of ten, she will virtually coax you to the whelping area or bring a sick pup to you—or if it has died, simply push or carry it outside. When the bitch and/or the litter are ready for handling or simple inspection, you will most often be extended an invitation to see them—usually after they begin distinguishing light patterns.

This is not to say that each of the pups should not be examined and monitored daily for signs of wansing health or neonatal injury. At Rudelhaus, we check and weigh each of the pups regularly, but they are not forcibly weaned, nor do we do any unnecessary cleaning of the sheds or move the puppies from one area to another unless absolutely necessary.

If this sounds like neglect, it is not. Part of the canine mothering process is to clean up every bit of discharge, fecal matter, etc., from the shed, and it is physically good for them. Were we to try to do this for them, we would not only create an unnecessary worry for the bitch, but would also prevent a natural physical process which, while seemingly detrimental, is physically most beneficial to the bitch. Generally, when the pups are beginning to eat from their mother's dish, when the bitch has openly pushed them out for inspection, she is ready for you to take over the ''house cleaning'' too.

CHAPTER IX. PUPPY DEVELOPMENT

THE FIRST SIX WEEKS

The number of puppies in a litter depends on many factors: the age of the bitch, whether she has had any previous litters, her health and the hereditary factors involved, and a good dietary and immunization program prior to and during the pregnancy. A bitch that is allowed to conceive in her first heat is likely to throw a very small litter and the

Banzai (male) at 8 weeks. Reported 75% British Columbian/Eastern Timberwolf/Alaskan Malamute. Photo: Linda von Hanneken.

45

puppies may be fairly small. A more mature bitch, in her second heat cycle or later, will likely have from five to as many as fourteen pups, although the latter number is uncommon. Seven to eight puppies is a more average figure for litter size.

Newborn puppies average five to six ounces, depending on the size of their parents. Their skin is tight and the hair is quite short. The ears are bent over and do not begin to stand up until they are at least five to ten weeks old. They can do little but mound together for warmth or nurse for the first three to four weeks.

At ten to fourteen days, the puppies' eyes open, although they will not see clearly until they are about five weeks of age—especially in bright sunlight. As they open, the eyes are dark and have a bluish cast to them. As this bluish cast begins to disappear, the vision gradually becomes sharper. I am often asked, "when do the eyes turn yellow?" They may never turn yellow. If the puppy has inherited the dominant genes for eye color from the dog side of it, the eyes may range from amber to dark brown, or to blue if some of the lineage is from Siberian Huskies. If it has inherited the dominant wolf gene for this particular characteristic, the eyes in the adult may range from a clear lemon color to amber, and this again, being a gradual process, may not be complete until the animal has reached maturity.

Each day you look at the puppies, the colors and markings will change slightly and the hair will gradually become longer. The "puppy colors" do not become set for nearly two months. Indeed, one can not be certain that the colors and markings exhibited by even a 6 month old pup will be those it will exhibit in its adulthood, as it may change color altogether by the time it is a year old. From the pups I have observed as they grew up, it seems that a pup will lighten rather than darken as it grows, though the masks and other markings may become more prominent.

At three to four weeks, the pups are moving around some and gradually begin to play and "rough-house," though they have little coordination or balance. The more dominant and self-confident pups begin straying a few feet from the security of the litter and their mother as they begin to see more clearly. Many will even begin taking food from their mother's bowl and tentatively seek or accept human contact at four to five weeks; but again, this depends on the human's relationship with the bitch.

It is at about three weeks of age that I personally feel that human contact and socialization is most effective, still without removing them

Sasquatch and Shadow, reported 73%. Photo: Corienne Cherry.

from the bitch's care. If the bitch is accepting and trusting of the human, the pups will learn to be also, by following her example. They also learn very quickly where the food comes from, who is the primary babysitter, and "where mom goes for love and affection." This same rationale works in terms of training. (See Chapter XI).

SOCIALIZATION AND WEANING

There are many breeders who feel that you should begin "socializing" or handling the pups from day one. Others feel that they should be taken from the bitch and the litter and bottle fed before their eyes are opened—which means LONG before they are naturally weaned by the bitch. There are several reasons why I do not recommend this, though I have met many wolves and Wolf Hybrids which have been separated from their mother this early and who have fared well.

The reasons are these:

1. Pups who are unnaturally and prematurely weaned do not receive the natural benefits of the colostrum nor the benefit of the maternal

antibodies to disease received through their mothers' milk. Physically, this is of greatest importance.

In addition, more handling than the mother gives them in the first few days of their life is not good for any species. From a physiological point of view, their bodies are still very fragile and their bones and organs not fully developed. They need even warmth and lots of sleep. The cushion of warmth provided by the litter affords this naturally. The often (unintentionally) rough human handling, compared to the mound of soft puppies is sometimes more than they can take. Add to this the fact that the viruses which cause most canine diseases may be brought to the puppies by the breeder himself on his clothing or shoes, and you can easily see the reason for this opinion.

There is still no way of determining when the maternal antibodies cease to affect the pups. A puppy weaned at seven days of age is going to be pretty defenseless against distemper, parvo, corona virus, etc., unless a strict and continuous program of vaccination is begun almost

Cody at 3 months, reported Tundra/Timber/Husky. Photo: Onica Gilmore.

Banzai (male) at 9 weeks.
Reported 75% British
Columbian/Eastern Timberwolf/
Alaskan Malamute. Photo:
Linda von Hanneken.

immediately. The vaccinations will do no good until the maternal antibodies wear off, but given the fact there there is no way of determining precisely when this will occur, the prematurely weaned puppy is in serious jeopardy if one waits the normal five to six weeks to begin innoculations.

2. It has been well documented that puppies fed exclusively on canine milk replacer have a high probability of developing juvenile cataracts. When pups which develop such cataracts are returned to a complete diet (i.e., nursing from the mother), the cataracts are generally reabsorbed. There is some evidence that the addition of glucose and fructose to the milk replacer helps deter the condition but it is still incompletely understood how or how reliably. (See Appendix C).

3. Pups removed from their mothers and their litter at an unnaturally early age do not receive the psychological reinforcement

provided by the nurturing of the mother and the comfort of the litter situation. They do not have the chance to learn and develop socially, either with other animals or in a way that is not totally manipulated by humans, and as they mature may (just as human children), "revolt" against such manipulation.

From a social point of view, the animals which I have observed that were prematurely or forcibly weaned exhibited one or more of the following undesirable characteristics, while their littermates who were left with their mothers until naturally weaned a few weeks later, did not:

Fearful and distrustful attitudes and the tendency to run and hide from people or new situations.

And/or aggressiveness which, when confronted even playfully, deteriorated into fearfulness and hiding.

Acceptance of only one or several of a family but fearful of or aggressive to outsiders.

You will note the repetition of the word "fearful." A pup's confidence is built upon security. The bitch provides this security initially by the warmth of her body and the natural nourishing. As the pup begins to exhibit its natural inquisitiveness, she is still near enough 24 hours a day to "rescue" and comfort it. She will wean it from a total nursing diet at the appropriate time while continuing to provide the security it needs when possible when l) she feels confidence that it will be cared for; 2) when there is no threat to its health or security; 3) when she feels that it is "ready for the world"; or very lastly, 4) when she is tired of caring for it. I mean to emphasize the last because some dogs nurse their young long past the time when the pups could and should be eating solid food if the first problems do not seem to be satisfied. A bitch may run off its heretofore trusted owner if she does not feel that it is yet safe to "turn the pups loose."

SOCIALIZATION WITHOUT WEANING

A good socialization technique that should begin the first week or two is to lift and cradle the cub for a minute or two several times each day. This helps the pup to accept its role with the owner without fear. Putting it down after a very brief time helps to reassure it and the bitch that no harm is intended. As the puppy grows older and its eyes open it will be comfortable with the smell of your body, and the petting and as it sees the mother interacting with you, will adopt the same behaviors without fear.

By listening very closely to the bitch, you can learn to mimic the high pitched whine by which she calls the puppies to her, or develop your own method of calling them. Whenever you feed them, first start calling them to you—"Kid! Kid! Kid" (or whatever—so long as it is the same call each time) so that they begin associating you and the call with the arrival of the food and affection. They'll soon come streaming out each time you call.

NATURAL WEANING

The cubs will nurse for as long as the bitch will permit, but they will start eating from her bowl of softened food at about three weeks of age. They should be allowed to nurse for as long as she will allow it to avert the possibility of diarrhea caused by a sudden change in diet, but as soon as they begin "stealing" her food, they should be provided the same food (softened) in a separate area so that they can begin getting on a solid diet and the bitch will get her rations.

Although initially, a good portion of the food will end up ON the pups instead of inside them, they very quickly learn to overcome the problem and a large bowl set before them very quickly disappears. We do supervise the eating, however, as the more aggressive pups will climb right into the bowl, pushing the less agressive ones from the food if allowed to.

We continue to use a good puppy chow, softened and mixed with baby rice cereal, powdered milk and calcium and general vitamin supplements, feeding several times a day. As their digestive systems develop more fully, the mixtures can gradually be thickened. There should *always* be fresh water available to them.

The bitch is more than happy to let you do the work of caring for her litter now, although she will be closely watching to make sure that they are not threatened. They are nearing the age when they can be removed from the mother altogether.

Complete weaning can be done at five weeks of age, although like other dogs, six weeks would be more desirable.

CHAPTER X. PUPPY BEHAVIORS

Social development and play activity in the litter and the relationship a puppy has with its mother will determine to a great extent, its adult personality. Puppies do little besides eat and sleep for the first two or three weeks after birth, but even at this early age, dominant pups compete successfully against their littermates for milk.

The more dominant pups will be the first to begin to leave the security of the mother and explore new territory. It is a dominant instinct that leads them to do this, but it is a submissive instinct that causes them to turn back each time to the safety of their mother or the whelping shed. A puppy's life is a constant state of flux between dominant and submissive instincts.

Like wolves in the wild, adult Hybrids "teach" the puppies by example, from which the puppies learn to mimic advantageous behaviors such as those used for hunting and reproducing. Of course, puppies also learn to mimic human behavior. If a person does not wish a cub to behave in a certain way, he could reduce the occurrence of that behavior by not behaving that way himself.

Playtime is a chance to learn and develop, both physically and socially. The pups are at first quite clumsy, falling over their littermates and their own feet and as they begin to try short little runs, often ending up turned around rear end first. Each new element of their environment that they discover will send them whirling back to the whelping area and it is comical to see them turn tail as a leaf or tiny twig drops into the yard.

Rough-housing and tugs-of-war are constant play activities, as their coordination, skill and social development sharpen. Skills shown during play are the ones puppies will use later for survival.

Many people interpret sexual mounting at this or any other stage of

Rebel—reported 50% wolf/white Shepherd. Photo: JoAnn Hubbard.

development to be purely a sexual urge. In wolves, mounting is also exhibited as a display of dominance, by both males and females, and this is also evidenced in puppy behavior. By observing the puppies' play activities, it can easily be determined which pups will be more dominant and which more submissive, all other circumstances being equal.

In addition to the above, play behavior includes attention-getting and hunting activities such as stalking, pouncing and digging often quite deep holes in the ground. As most Wolf Hybrid owners can attest, a person strolling through an animal compound who does not keep watch on the ground often ends up with injured ankles. The pups will dig furiously for several minutes, then abandon it for another game.

A puppy isolated from littermates or other puppies during this growth period may have difficulty in developing these skills and attaining the confidence needed to interact with other animals as it ages. If removed too early or too late, excessive fear or aggressiveness may result. Leaving the mother and the litter at the right time, and gradual weaning, reinforce the dominant instinct of confidence a puppy needs to fend for itself and reduce its psychological dependence. A puppy benefits most from being introduced to a new home shortly after weaning. While puppies learn to interact positively on their own with people anytime between three and fourteen weeks, five to six weeks is probably the best time for introduction to their new surroundings.

CHAPTER XI. EARLY TRAINING

The training of the puppy begins with the ride home from the breeder. While the puppy needs love, warmth and positive physical contact and reinforcement from its new owners, it should not be held all the way home. The puppy may whine and whimper, but the owner should try to relax and ignore it. As it grows up, one will not be able to cuddle and hold the animal each time it whines, and it is best that it not learn that this attention-seeking behavior will bring rewards. Better to

Cree at 9 weeks. Reported 75% Mackenzie Valley/German Shepherd. Photo: Ann Ballard.

pet and comfort the puppy every so often, and then retreat.

The puppy may suffer from motion sickness or have to eliminate, and the new owner should be prepared for this and not get upset if it happens. Motion sickness can be somewhat reduced if the puppy has not eaten for several hours prior to the trip. The simplest alternative to both the problems of fear and the probability of a mess in the car is to take a kennel carrier with a rug in the bottom and cover it somewhat so that it is darker (and safer) to the pup.

On arrival at its new home, the puppy should have an opportunity to explore the entire house on its own. This gives it an opportunity to learn to find the "safe" places and also reinforces its independent instincts. Show the puppy its food and water bowls in the place where it will always be fed and offer a blanket, rags, kennel carrier, cutout cardboard box or similar arrangement for sleeping. Unless taught differently, the puppy will most likely adopt a spot under a couch or chair where it feels hidden and safe. This at first might seem fine to the new owner, until the damage due to chewing on the furniture is discovered. The puppy doesn't know any better. He considers the underside of that particular piece of furniture to be the top of his "den" and therefore, his own.

Whether it sleeps in the bedroom or somewhere else, the puppy may whine, cry or howl the first night or two. It misses its mother and littermates, is afraid, and wants to be comforted and close to its owners at all times. Getting to sleep (on the floor) in the bedroom can give it a great feeling of security. Isolation in a room apart from the owners may cause a fear of isolation and cause it to continue to act this way at night indefinitely. Some people feel that placing a loud-ticking clock near the pup will help to lull it to sleep. The best way to teach a Hybrid to overcome its fears, however, is by its living through the fear-creating experience and being praised in the morning for its efforts.

HOUSEBREAKING

Although there are no hard and fast rules for housebreaking, it is generally easier and faster to train the puppy to go outdoors from the start. Newspaper is too much like something to chew up to the cub, and paper training only condones eliminating indoors—a habit the animal which must then be taught is bad at a later time.

New owners with other pets face an additional challenge with

Hybrids in that, though they understand they are *supposed* to go outdoors, they feel compelled to "mark their territory." Understanding this may help the frustrated new owner when after urinating outside, the animal calmly walks back into the house and does it again. Physical punishment will not help, nor rubbing his nose in it. In fact, it may make things worse. Scent marking simply is not the same as not being "housebroken." It is better to immediately scold the pup in a shaming way, pick it up and take it back outside to its regular area to try again.

Always praise him when he does urinate and eliminate outside to reinforce his desire to wait whenever possible to go outside to do it. Remember, a Hybrid, more than other dogs, will try its best to do what will please its owner.

Owners should establish regular feeding times and stick to them. Feeding may be several times daily until three months of age and then reduced to two times daily until the pup is about eight months of age. Since eating usually stimulates the waste elimination reflexes, a regular feeding schedule also helps to regulate toilet training. Immediately after each feeding, take the puppy to its toilet area and wait patiently for results. The puppy will need to go out each time it awakens too. Remember, all puppies have difficulty in "holding it in" until they can get your attention because their digestive tract is still developing. It is important not to scold "mistakes" and not to rub the puppy's nose in its own wastes because such actions force extreme submission on the pup. Instead, each time you catch him in the act, snatch him up and scold him and take him outside to complete the job. Then give praise and petting for each success.

Almost all puppies could be housebroken within one or two weeks if handled and observed properly. If the new pup has not been housetrained within a few weeks, and is not suffering from health problems (such as diarrhea or urinary infection), it is probably the owner's inconsistency or misunderstanding that has caused the delay. The problem of housebreaking is more often one of training the owner to be receptive to the needs of a tiny, not yet fully developed animal with a very limited warning time and less of training the animal to try desperately to hold it in until the owner understands its signals.

Perhaps the easiest housetraining aid is the installation of a dog door. In the whelping shed, the bitch keeps the entire shed free from fecal matter and the pups learn by her example at a very early age to go outside to eliminate, if only a few feet. This attitude is transferred to the owner's house when they are brought in; but if they cannot get outside immediately, housebreaking problems begin. The dog door allows them to go out as they need to without having to wait for the owner to recognize their need. With training to the dog door, which at Rudelhaus, happens by imitation and following of the mother, we have had as many as 19 "unhousebroken" cubs introduced to the house simultaneously with few accidents in the house.

CHAPTER XII. TRAINING A WOLF HYBRID IS DIFFERENT

"Training" of the Wolf Hybrid can be very frustrating to new owners. It is often more appropriate to say that the Hybrid trains the owners in that the animals may learn very quickly and easily the behaviors that are expected of them, but will perform them "if they feel like it." It is perhaps more accurate to say "if they feel it is appropriate." "Training," to me, means both the curbing or elimination of unwanted behavior and the teaching of new, desired behavior, and for that reason this section is not restricted solely to formal training, as dog owners usually think of it.

Hybrids absolutely need love and affection and praise for good actions, firm but gentle disapproval for bad actions. They are loving, loyal animals and their instinct and need is to be an integral and active part of the family that adopts them. They will want to be near you whenever possible and will try to be in the same room with you most of the time. Particularly because of their size when grown, they should be trained from the start to stay off the furniture and not to jump up on you for attention. At one year of age, a Hybrid can easily knock you to the ground in play.

A Hybrid is initially shy and it will be best for it and the owner to gently introduce it to the new house and surroundings without lots of commotion. Unless you are providing a companion animal for it, it will spend some time hiding and sneaking out to explore its surroundings a bit at a time. Introduction to friends and non-family members should be gradual and quiet, allowing the pup to develop confidence to deal with each new experience. Don't allow anyone to spook or tease the pup. It is, after all, like a baby and has fears just like any small child.

Play with children should be well supervised so that the animal does not develop a fear of children who have treated it roughly, and so that the children are not nipped in play. A child plays by running, poking, hitting, or throwing things. Up until now, the puppy has only been playing with its littermates and pouncing, chewing, biting and scratching have been a part of that play. Obviously, for the benefit of both, the play behaviors of both child and animal must be modified. This may take some time, and a lot of supervision to teach them that this is no longer acceptable. Young children can cause many problems for a new puppy and they should be prevented from handling him roughly. Introduction to a child should be done when the puppy is quite young so that strong, loving bonds can develop and future problems avoided.

If the puppy is to be raised with other animals, it should be introduced to them from the start. Hybrids can be raised successfully with cats and other animals, provided the puppy is the submissive one from the start. It is simply a matter of the older animals "being the boss." It's their territory.

You should provide the puppy with toys, shoes and rags that it can chew on and toss about, particularly while it is teething, for it will want and need to chew on something. If allowed, it will quickly begin demolishing the furniture, boxes, papers, books and anything else it can get its teeth on, another habit which will have to be broken. Better that it is taught through praise and enjoyment to chew on its own possessions than yours.

REWARD VS. PUNISHMENT

One of the first things you will realize about the Wolf Hybrid is that normal methods of dog training do not work at all. If you have been fortunate enough to hear the quiet, high pitched sound the bitch uses to call her pups, you can imitate it, alternating that sound with calling in your normal voice, to train the cub to come to you, gradually eliminating the high pitched whine.

Just as the puppy learns to eat solid food by imitating its mother and stealing her food, so does it learn other behaviors by imitation of its mother or its owners. A domestic dog may learn behaviors to please you or out of fear of you. A Wolf Hybrid learns behaviors because you, in effect, have become the alpha, the leader of the pack, and it *must* please you or earn the humiliation of your disapproval and disdain.

If you watch a group of wolves or Hybrids, you will notice that there is one male leader and generally, one female leader (the alphas or dominant animals). The rest of the group will go to outrageous lengths to show their love and submission to the alphas. They are kept in this place, sometimes by aggression on the part of the alpha, but more often by shaming and disdain. It is this same principle that Wolf Hybrid owners use in their training—lavish praise for good and acceptable behavior; shame and disdain for unacceptable behavior. Just as they react to a myriad of vocal pitches and body language among themselves, so do they interact with humans. A shaming can be absolutely unbearable to a wolf or Wolf Hybrid. Physical discipline, on the other hand, teaches them to react in kind.

Some people disagree on the merits of reward vs. punishment in training the Wolf Hybrid. There are times when MINIMAL forms of punishment are required, but reward for positive behaviors is the most consistent means of encouraging repetition. Some Hybrids would sell their souls for tidbits of food and praise and many people train their cubs by trading good behavior for food. This should be minimized, however, and praise substituted for treats.

When punishment is required, probably the method most easily acceptable to the pup is to discipline it in the same manner its mother would. Pushing it sharply to the ground on its back and pinning it by the throat and growling usually gets the message across. Some owners go so far as to bite the animal on the nose for particularly bad behavior.

REINFORCEMENT

Like a wolf cub in the wild, the Hybrid cub begins learning before it even opens its eyes. From birth, they are mounded together for warmth. If they start straying from the mound, the mother pushes them back. If they stray too far from the den, the mother patiently hauls them back.

Teaching simple commands can begin on the first day in the new home; however, it takes time for the pup to learn these commands, and its attention span is very short. It is wise to be patient and to praise any successful response to a command. When a Hybrid has learned to obey as a puppy, it will be more inclined to obey when problem situations occur.

The puppy will begin to respond to its name very quickly if it is called gently, teasingly, cajolingly, and praised and petted, and perhaps fed a tiny tidbit of food each time it does respond. When its name is part of each interraction with the puppy, it will soon respond to it in one way or another.

The commands to come and to sit are probably the most easily mastered. When you call its name and say "come," whether for feeding or for play, it will learn very quickly to come bounding to you by association with pleasant things.

Except for feeding time, a puppy will very often bound up to you and plunk down in front of you when you do not immediately play with it. If it is simultaneously told to sit and rewarded for doing so, this too becomes reinforced as a good behavior and will eventually begin to sit on command. While this may also involve a little bit of physical manipulation if the pup doesn't accidentally do it all by itself the first few times and will be of a very short duration, it doesn't take long for the association of "sit" and praise or food to sink in.

The command to lay down is merely an extension of "sit." After the puppy has learned to sit on command, have him do so and pet and praise him. Then gently draw his front feet out to a laying position, while saying "lay down." He will at first resist this movement and jump right back up, but with gentle persistence and lots of praise when he stays in the position (even with your help) he will begin to get the idea.

These are the simple commands that the puppy can learn when still very young. One must remember though, that its attention span is very, very short and although it may learn the commands, may stay in the desired position for only a very short period of time. Later, as it begins to mature, more can be expected of it, but the basic training is already in place.

"Stay" is pretty tough, even for a mature Hybrid. Hybrids are easily distracted, and though they may do quite well for a moment or two, if distracted, or if, as some people are want to say, they decide they've had enough of this training stuff, may simply get up and wander away, or chase a shadow, or bound up for their reward. A hand signal, such as the signal used by a traffic control officer for "stop," used with the word "stay" usually works better to introduce them to the meaning of the word. They are so focused on the hand position (which has probably also been used to deter destructive behavior) that they temporarily forget the fact that you may be moving away from them. Still, the "stay" will only last for a few seconds, gradually becoming longer as the pup understands that this is what it is supposed to do and that it will be praised and rewarded for what it has done at the end.

LEASH TRAINING

All dogs should be taught to walk on and be responsive to the leash. There will be many occasions throughout their lives when use of the leash will be necessary, such as visits to the veterinarian or when others want to meet the sometimes over-exhuberant animal, if on no other occasions.

The puppy should become accustomed to wearing a collar as early as possible. The collar must be checked every few days to be sure that it is loose enough not to be binding, but not so loose he can slip it off at will or get his feet or teeth tangled in it.

Between ten and twelve weeks of age, a very light-weight leash should be attached, allowing the puppy to "take the owner for a walk." The first few ventures with the leash are for getting the puppy used to something new and not for training.

I favor a very light-weight chain slip (choke) collar, not so much for their control value, but because when actual training begins the dog associates the sound of the chain slipping through the rings with movements desired of him. A small clip can be attached to the chain to prevent actual choking potential of the collar, which can be adjusted as the animal grows, while still allowing the slipping noise of the chain during training.

I will not attempt to present a complete training manual here. There are many good books available written solely for that purpose. The point in recommending the slip collar is that the sound of the chain slipping through the rings very quickly becomes an advance signal to the animal that "we're going to change direction," "we're going to stop," "You're going too fast," without need of any physical direction from the trainer. Again, I would emphasize that gentle, persistent consistency in training is necessary, along with enormous amounts of praise for each good performance, even if momentary, and an understanding of the initial timidity toward each new action required.

Training sessions should be short but regular. The animal may work very well one day and very poorly the next. It will always be subject to distraction, even as it gets older, and while it may do everything it is asked to on the leash, off the leash can be an entirely different story.

After, or as a break in training, lots of affection and praise is appropriate and a time for joyous romping—a welcome relief to any animal after even a short training session—will come to be anticipated as another reward.

CHAPTER XIII. HOWLING

Depending on the location where you live, howling may be either a problem or a joy. The owner may take extreme delight in the "singing," but his neighbors may not—particularly if it occurs at 3:00 in the morning.

A puppy's first attempts at howling are somewhat comical, most often coming in the form of high pitched "yips." A puppy's "whine" or whimper is high pitched, but obviously more frantic than the attempted howl. Its bark is much deeper and in pups, usually heard in play situations. Its sounds of pain are unmistakable. Its attempted howls are funny. It takes a pup time to develop a full throated howl, even when retained with the litter or other adult Hybrids.

Wolves and Wolf Hybrids typically "sing" at dawn and just before feeding. The howl is a form of communication between wolves, and with time, the owner will begin to differentiate between howls which begin for different reasons, be they triggered by high-pitched sounds such as sirens, by people or animals approaching their territory, of a warning nature or the spontaneous and very beautiful low, mournful howl most often reproduced in movies. Wolves will spontaneously howl more often when there are other wolves in close proximity. They will also howl more often at night under a full moon. Contrary to all the mythology which ties the wolf howl to the moon, it is probable that this occurs because there is more light and thus more visibility and reduced sleeping hours, rather than any other direct tie to the phases of the moon. This probably also accounts for the pre-dawn "singing" and the origination of the "hour of the wolf"—the short period of time from first light till dawn.

I do not feel that howling should be discouraged unless it becomes excessive and a problem with neighbors. I am not really sure how one

Cody—reported Tundra/Timber/Husky. Photo: Onica Gilmore.

could prevent howling. It is instinctive and enchanting and when all things are considered, is part of the reason that many people buy Wolf Hybrids to begin with—a very perceptable part of the image of the wolf itself. Perhaps it is better to say that people buying Wolf Hybrids should anticipate it and be in a position where it will not be a problem with their neighbors.

CHAPTER XIV.
GUARD YOUR POSSESSIONS

As mentioned before, all puppies chew things up. It's not only fun and normal puppy behavior, but, as in children, it helps in teething. This chewing activity continues all through puberty and into adulthood and, as mentioned before, Hybrids mature much later than other dogs. A wolf or Wolf Hybrid cannot be considered to be mature until it is at least two years old, and juvenile instincts and behaviors are not lost instantly when it passes a certain age.

Hybrids are natural born thieves and, if they see something that looks interesting to them, they will grab it and run, usually hoping that you'll chase them. The next time you see that favorite pair of shoes, your check book or even your living room rug, it may be in tiny pieces scattered all over your yard. They seem to have an intuition for what things are most valuable to you and will forsake all normally interesting things for those special items. They not only chew, but devour, and one is often rather astounded at what is found when cleaning the animals' yards; often, if the owner is not particularly watchful, shreds of money. Even in adulthood, we have occasional "sneak attacks" on our possessions and long ago learned the hard way that nothing not totally out of reach or behind securely latched doors or fences was entirely safe.

We had one young male who was strongly attracted by my daughter's stuffed animals and though very well behaved in every other way, would steal them, take them outside and bury them whenever he could. We would find them daily, with just a snitch of them sticking out of the ground. Finally tiring of that game, my daughter gave him his own private teddy bear, and from that time until he was nearly ten months of age, he carried it with him wherever he went like a security blanket and left her other stuffed toys alone. One day, for no apparent reason, he tore it to shreds and never stole another toy after that. Others we've owned have continued to steal through their entire lives.

Hybrids are, in comparison to other dog breeds, incredibly intuitive and cunning. They seem to know when you are becoming complacent and trusting that "they would do no wrong" and then turn around and prove you altogether wrong. They learn to open gates that are not securely locked, cupboards, even refrigerators. An animal that has heretofore been perfectly content to ride in a car and for a time be confined to a car or house, may suddenly totally disintegrate its interior. An animal in a pen with no "ground wire" that has never dug before can, within five minutes dig an escape route and be gone before you even realize what it was thinking about.

I would hate to have to make an honest listing of how much money has been lost to my wonderful thieves, both in terms of articles that they have torn up and eaten and in terms of actual money eaten. It would be embarrassing. But it is not so much the fact that the animals are being "destructive" as the fact that chewing, tearing, shredding, scattering and yes, ingesting of fun-looking things is common and the owner should be expecting it and HIDE the things he wants to preserve, just as you hide important or dangerous things from a very young child. You can not, under any circumstances, presume that an animal (or child) will not get into or eat anything, no matter how bad its taste. And you certainly cannot presume that any a puppy of any breed will be able to perceive that any particular object in "its territory" is supposed to be and must be left alone. Sealed containers of any description are no exception. Cigarette lighters (which are potentially highly volatile) metal cans, plastic containers, chairs, couches, rugs...they are all fair game. To the unitiated, this may seem like an exaggeration, but I have watched attempts to drag my wall-to-wall carpet out through the dog door!

Aside from the protection of one's belongings, the problem also is one of protecting the animals from the things that they steal. Poisonous chemicals, of course, are fairly obvious hazards. Soft plastics, such as bread bags, plastics or aluminum wraps, socks (a favorite), can become blockages in the intestinal tract. Hard plastics, bits of metal or thicker aluminum wrap, nails, and larger objects may become imbedded in the digestive tract, whether in the mouth, throat, stomach or the lower digestive tract, and can cause severe internal injury, internal bleeding, severe pain and death.

Search your area for anything potentially harmful to your animal and remove it. If YOU can move or remove it, you can assume that they can too.

CHAPTER XV. SOCIAL STRUCTURE
AND AGGRESSION

Much attention has been given to the social structure of the wolf pack in the wild. A wolf pack is a family group and its members are usually an alpha male and female, and several of their offspring or other related wolves. They do not necessarily remain together indefinitely, as some of the younger individuals will go off to mate and

Balrog—reported 100% Mackenzie Valley Timberwolf. Owner: Laura & Fred Kerns; Photo: Rose DeProspero.

form a new pack on another territory. There is generally no more than the one litter produced by the alpha pair each year.

There is much literature available about pure wolves and the pack social structure and I would strongly suggest seeking out the available published literature[8].

It has until recently been generally accepted as fact that wolves in the wild mate for life. However, there have been exceptions observed, particularly in cases where one of the mates has died. It has also been speculated that one reason only the alpha pair breeds each year is directly related to available food supply.

Wolves that leave the pack to mate and subsequently begin a new pack face a great challenge. They must move to and defend a new territory and any individual of another pack that accidentally strays onto that territory must be successfully challenged.

Within a pack, there is the well-known "pecking order," particularly with regard to mating. The dominant-submissive posture may break down, however, in certain situations, such as when a younger, stronger wolf assumes the dominant role in hunting. The "beta" animals will in most other situations, however, always remain in a submissive role to the older, dominating "alpha," until they either leave leave the pack or defeat the dominant wolf in a battle, which either kills the formerly dominant wolf or drives it from the social structure of the pack.

The significance of this to a wolf or Wolf Hybrid owner or breeder is at least two-fold. In the case of the captive pure wolf, with an abundance of readily available food, based on this criteria, the theory that only one pair within a pack will mate does not hold up. Likewise, in captive and particularly domestic situations, pure wolves are often rather promiscuous from year to year and while they may prefer the company of one particular mate to another, are often bred to others and then returned to their mate after the period of fertility is over.

With the addition of the dog blood in the Hybrid, the animals may be much more unselective, and care must be taken to insure that breeding is strictly limited to breeding between the selected pair so that no other accidental breedings occur.

AGGRESSION

One very important problem (sometimes disastrously so) that doesn't seem to occur to many new owners is that the personality may change drastically as they "come of age." During breeding season

Moon Dancer—reported 5/8 wolf (male). Owner: Jean McCarthy.

and until the pups are several months old, unless the animals are separated from others of their own sex, terrible, bloody battles may ensue, between females, between males, between parents and children who are apt to be viewed as challengers to the "alphas." I've seen worried, jealous females trying to pull puppies from another litter through chain link fence where the pairs and their litters were in adjoining enclosures. I've watched fathers trying to kill rather submissive sons, and mothers chase after daughters and formerly compatible sisters.

By the preceding warning, I do not mean to imply that the Wolf Hybrid is unusual in this respect. Whenever one is dealing with dogs during mating season, one must be especially sensitive to an alteration in the reactions of the animals and take precautions accordingly. In the wild, an unsuccessful challenging animal has the prerogative of running off and escaping. In a captive situation, however, where there are no avenues of escape, an artificial situation of "do or die" has been created.

There are no means of escape. This applies to Wolf Hybrids as well as to pure wolves. If the human has seen fit to have the animals in an artificial environment, then he must take care to protect those animals from the situation in which they have been placed.

HOW DO YOU DEAL WITH AN AGGRESSIVE SITUATION

Humans who jump in and try to break up such fights usually encounter one or the other extreme results. They may find themselves in the perceived position of being a challenger themselves and suffer severe injuries for their attempts to help the "underdog," or may find that they cannot even get the attention of either animal, by whatever means, but still may accidentally end up being injured without the animal ever even realizing that they were there.

What should one do in the event of serious confrontations between their animals? Erich Klinghammer of Wolf Park[9] keeps fire extinguishers handy for preventing or breaking up confrontations between the wolves and the humans. At Mission Wolf[10], a very loud gun is shot into the air, which generally breaks the focus of the animals' attention and allows a very short amount of time to separate the animals. Zoos and many wildlife exhibits usually use high pressure water hoses, although I have tried this in -15° weather with little success until the animals had very nearly exhausted themselves and were going already into shock.

I have never been intentionally bitten by any of the animals I have raised from puppies, but I know of several who have, mainly by pure wolves or very high percentage Hybrids, and for the most part, because they violated the "social code," as wolves understand it. If you suddenly jump in to disrupt the established hierarchy, you must expect to suffer the consequences. Although this doesn't happen in every situation, the potential is there—particularly with pure wolves,

F^1 and F^2 crosses.

While most confrontations really involve a one-on-one battle, where more than two animals are confined in the same area, it is common to see the entire group become involved—generally on the side of the more dominant animal. The ''pack'' joins in, biting and pulling on the tail, the flanks or wherever they can grab hold, to the detriment of the poor ''underdog'' who cannot escape from the contained area.

Most owners of lower percentage animals will never encounter such a situation, particularly if they will take the care to prevent such an occurrence by separating the animals into pairs before they become of an age where such confrontations begin to happen.

Owners are apt to take such confrontations personally, and I remember in the early years my feelings that the animals I loved so much were demanding that I choose one over the other and my refusing to do so. This is a very naive, uneducated and uncomprehending attitude on the part of the owner, but a very common one. Whether any one of them is your "favorite" has very little to do with a wolf or Wolf Hybrid's perception of its pack structure and how he or she deals with it. If you have more than one pair, you must expect it (whether it happens or not) and be prepared to deal with it.

Oljato, reported 77%. Photo: Bobbie Holladay.

73

Persecution by the pack is not confined to social or territorial challenges. When an animal is hurt or sick, it often becomes the subject of severe abuse from the pack. It is wise to separate an injured or ill animal from the rest of the animals until it is once again in good physical condition. I don't mean to imply that it should be "isolated" unless it has a contagious disease, but protected by a fence which will prevent injury from the rest of "the pack."

Perhaps the most important things to understand are prevention of such incidents and refraining from interfering if a human's safety is not involved, including your own.

In many cases, the confrontations between the animals will be simply reinforcement of dominance by the alpha and no serious injuries will occur so long as no other influences (such as human intervention) occur. But each animal will react differently, and sometimes the "underdog" will be the one to attack and refuse to quit and attack the person attempting to help it, not necessarily because the person is trying to intervene, but because its attention is suddenly and temporarily diverted to the human, rather than the animal it was fighting with. On the other hand, once a physical confrontation has taken place between the human and Hybrid or wolf in this sort of situation, it is very often difficult to reverse the damage done to the personal relationship between the two. The human now distrusts the wolf or Hybrid; the wolf or Hybrid also may now distrust the human. The most common result is that the animal is given away or put to sleep.

It is now the human which must undergo a "metamorphisis." The human has the capability of attributing all sorts of personal hostility on the part of the animal, which most likely doesn't exist at all, but when the human is hurt, it often makes no difference. The bond of trust must be re-established. The animal has (as the phrase goes) bonded to a particular human or humans. Although the animal *may* establish a trusting relationship with a subsequent owner, it would be far better for both the original owner and the animal if the bond could be re-established between the two.

An aura of fear in a human is easily recognized by another human. It is even more obviously sensed by an animal, and fear in any other often triggers the prey instinct. Another human who is not afraid of the animal may establish a very good rapport with it. For this reason, I would hope that anyone finding himself in a problem situation would at least explore the alternative of finding a good, suitable home for it before putting an animal down because it has been in a human/wolf (or human/Wolf Hybrid) confrontation.

With knowledge and understanding of the animal, owner/wolf or owner/Wolf Hybrid confrontations are the exception, rather than the rule, given the human's understanding and acceptance of the animal he is dealing with and his prevention of situations where confrontations might occur.

I have been talking about owner/animal relationships, but I would like to interject a thought about reactions of the animals to people with whom they are not in daily contact.

Wolves and Wolf Hybrids are very perceptive and are more trusting of strangers who are similar to the people to whom they have bonded. Just as they easily pick up on fear in humans, they are quick to perceive qualities and movements which they interpret as furtive or threatening and nervously react accordingly. It is, therefore, wise to evaluate their behavior and reactions to strangers from the other side of a protective fence before introducing them to the newcomers directly.

CHAPTER XVI. NUTRITION

There is much controversy among owners of wolves and Wolf Hybrids of any percentage about what sorts of diet the animals require. While most feel meat should be part of the diet, some feel that they require a total meat diet, and others feel that primarily kibble diets supplemented occasionally with meat are sufficient. There is even radical difference of opinion among zookeepers and preserves on this subject, with places such as Wolf Park[9] feeding strictly road kills and others, such as the Rio Grande Zoo (one of the holding facilities for the Mexican Wolf Recovery Program) feeding strictly kibble, supplemented once a week with bones "for the hassle factor"—the pulling, chewing and psychological needs of the animals. Because there is so much disagreement over nutritional needs, a description of what we feed at Rudelhaus is presented, with the balance of this chapter devoted to the reprinting of a long series of articles on nutrition published in *The Wolf Hybrid Times*.

As stated in Chapter IX, when our pups are beginning to seek solid food on their own, they are offered a high quality puppy chow which has been softened with powdered milk and baby rice cereal mixed with warm water to about the thickness of tomato paste at least three times a day.

It is important not to overfeed. Pups will eat until they throw up and as fast as possible, and can develop colic or canine bloat. The rice cereal, in addition to being very nutritious, is soothing to the walls of the intestinal tract and reduces the chances of developing diarrhea during the transition from bitch's milk to solid food. At about seven weeks, the puppy will probably be satisfied with the previous diet, along with a dietary supplement, such as K-Zyme or Pet Tabs and a little powdered milk.

Being part dog, we do not feel that the Hybrids necessarily need raw

Feeding time at Wolf Park, Battle Ground, Indiana.

meat, but do require more protein (via the dietary supplement) to grow at the optimum rate.

Most good quality kibbles are nutritionally balanced, and we feed one that is meat based, mixed with a gruel composed of canned dog food or soup stock, dietary supplement and water. We do recommend using a good formula of puppy food for animals under 1½ years because of the added supplements which puppies need for optimum growth—including the powdered milk it is generally dusted with.

It is peculiar to wolves that despite all the other non-nutritional things they eat with no ill effects (i.e., socks, scraps of paper, etc.), they have very sensitive digestive systems and change of diet can cause diarrhea, as will the initial change of environment to a new home.

Like children, an animal's behavior can be affected by its diet and whether or not its nutritional needs are being met, so proper diet is essential. Clean water, of course, should be available at all times.

NUTRITION

A Series Reprinted From *The Wolf Hybrid Times* (1987)

We all want to feed our animals "the best," and if we can't afford it, at least to provide what they need to keep them the healthiest, happiest animals possible. But what type of diet will best accomplish this? Does a wolf or Wolf Hybrid need a pure raw meat diet? They are, after all, descended from pure wild wolves who supposedly eat only raw meat. Or are the commercially marketed kibble feeds adequate? They have been developed by veterinary scientists and nutritionists specifically for canines. Or do these diets need to be supplemented by one or more of the various canned foods and supplemental vitamin products available? Is a "happy balance" a combination of each? If you feed raw meat, will your animals' personalities become more predatory and aggressive once having tasted raw meat or blood?

These are some of the questions that this series will attempt to address. It is recognized that WHT will not be able to fully and scientifically cover the subject—we are not animal nutritionists, nor will we recommend ANY particular commercial products. We will in this series attempt to present sufficient data to enable the reader to make an educated decision about the "best" diet for their animals.

The following table summarizes the response of readers to a questionnaire submitted to them.

Table 1

TYPE OF DIET

Kibble as the main part of the diet	93%
Raw meat only	7%
Kibble with meat supplement (often in the form of fresh road kills or butcher scraps	53%
Kibble with canned food supplement	9%
Vitamin supplement	25%
Table scraps supplement	24%
Rice supplement	4%

COMMERCIAL BRANDS
LISTED IN ORDER OF POPULARITY

Purina Hi-Pro
Iams Eukenuba
Purina Puppy Chow
ANFM
Wayne
Various other brands scoring very low

UNDERSTANDING THE INGREDIENTS

The first step in determining what commercially manufacturered food you want to use, if any, is to learn to read the "guaranteed ingredients." Reading one of these listings can be deceptive and confusing. In the first instance, imitation meats or flavorings add nothing to the nutritional value of the feed. They are purely and simply an artificial or imitation taste-tempting device. U.S. Government labeling requirements do not require that *any* of the flavoring ingredients actually be a part of the feed, i.e., that chicken flavored feed actually contain any chicken—only that the animal recognizes the taste as such. I'm sure we are all aware of artificial "simulated tastes and smells" from passing by fast food places. The incredible smoked and/or barbecue aromas we smell as we pass by don't necessary come from the meat being cooked inside, but are produced by a combination of chemicals which are vented outside to attract customers.

The array of colors, shapes and consistencies in which kibble comes has little to do with its content (with the exception of semi-moist vs. dry kibble) and these colors and shapes have nothing to do with the animals' needs or preferences. They have to do with the owner's visual perception of "better," for example, beef should be brown colored; chicken, yellowish. We don't like our kibble to reach home in a state of near powder due to the store's mis-handling; however, the canine couldn't care less and mixed with a little water, it mixes up to a tasty mash.

DEFINITIONS

If you buy a sack of dry dog food or a can, what are you actually getting? What does that list of ingredients actually mean? It is required, just as in "people food," that all ingredients be listed on the packaging in decreasing order of weight or percentage. The following are explanations of some of the terms used. Theoretically, at least, the contents must conform to the package label.

Flavor: As stated above, the source of the flavor must be recognizable to the pet, but the feed need not contain any of the named ingredient. But, if it is a canned food and is called "beef dinner" or "beef stew," for example, it must contain at least 25% of the meat product indicated on the label. If the terms "stew" or "dinner" aren't used, then the product must be at least 95% of the meat or foul named. According to an article in *Dog Fancy* (April 1987), "Other ingredients can be used in the product label if they contribute in some significant way to the product, even if they don't make up 25% of the contents, i.e., 'beef dinner with cheese' if it has at least 25% beef and enough cheese to affect the price or acceptance of the product."

Meat: For an ingredient to be classified as meat, it must be clean flesh derived from slaughtered animals and limited to certain types of muscle and flesh and can be only from cattle, pigs, sheep or goats, unless it is specifically referred to as another meat, such as "horsemeat."

By-products can include blood, lungs, kidneys, liver and other organs but cannot include hair, horns, teeth and hooves.

Meat Meal or Bone Meal are the dry or rendered products derived from "mammal tissue" with the same exclusions that apply to meat by-products.

Grains comprise the major component of dry feeds and may be cereal grains, wheat and oats, corn or many other by-products from cereal milling. Grains are the "primary sources of energy from carbohydrates," according to *Dog Fancy,* while protein is derived from the corn gluten and soybean derivatives.

Chemical Additives. The most confusing part of the labels are the dozens of chemicals which are listed. Chemical preservatives, such as ethoxyquin and "BHA" are sometimes shown in brackets as being preservatives; sources of certain vitamins are often indicated as being such. Not explained in any way on the labels are the chemicals or other "artificial stool hardeners" which are included in the product. This subject is one of many that will be covered in another part of this series, as well as the philosphies of different feed producers in the development of their feeds.

While the amounts of protein, fats, fiber and moisture are listed in general terms somewhere on the packaging, these percentages too are generally misleading.

The amount or lack of moisture in a product simply tells you if it is a dry product (about 12%), canned (about 75%) or semi-moist (18%-50%). (See product, label examples, Appendix D.)

The percentage of crude fiber present indicates the amount of the food that is not totally digestible and will not be utilized by the canine. The higher the percentage of crude fiber, the less will be utilized by the animal.

The crude protein percentage reflected is "measured by the amount of nitrogen present," not the quality of that protein.

According to *Dog Fancy,* the crude fat is "measured by the total content of the product that is soluble in ether. This ether extract includes neutral fats, fat-soluble vitamins and phospholids."

PROTEIN—WHAT IS IT?

In our survey, we found that most owners are principally concerned about the quantity of protein in their animals' diet and whether the source of this protein comes primarily from meat. Comparison of the content of most of the kibbles will show that the majority of the feeds derive the protein from vegetable and grain sources (see Appendix D) with the (to most people) indecipherable vitamins being supplied by chemicals. We have been taught for decades that protein is the single most important component of any animal's diet, from whatever source. Protein is essential to the tissue building process, and its assimilation by the body determines whether the body will be or become the "perfect body," an emaciated body, or an obese body.

Statements like the above are true, but are so broad and generalized as to mean absolutely nothing when looking at the protein levels provided in feeds. The problem is that the percentage of protein that is *available* as opposed to that which can be *utilized by* an animal in any particular diet is dependent on much more than the amount of protein required by the National Research Council. The amount of "usable protein" is based on its amino acid composition and the correct balance of the amino acids. Approximately 25 different amino acids are involved and must be balanced in a near perfect ratio for the protein to be of value to the animals. Ten amino acids, according to Robert Abady of the Robert Abady Food Company (also manufacturers of Nutra-Vet products) are termed essential and must be provided in the diet. The essential amino acids for the dog are arginine, histidine, isoleucine, leucine, phenylalnine, threonine, methionine, valine, lysine and tryptophan.

The second group of amino acids, termed "non-essentials," are glycine, alanine, serine, cystine, tyrosine, aspartic acid, glutamic acid, proline, hydroxyproline and citrulline, which Abady says are "physiologically as necessary as the essential amino acids in that life can not proceed smoothly without their presence in adequate amounts and in the proper distribution, although...they can be formed in the dog's body by an essential amino acid combining with other suitable chemicals." However, under physical stress, such as whelping and nursing and that of working and very active dogs, the essential acids are so depleted that the non-essentials are not provided by the body, and if not present in the diet, are unavailable to the dog. It follows then that the protein level available to the dog is reduced considerably.

"Additionally, because all essential amino acids must be absorbed at

Reported 87% female cub. Photo: Randal Bowen.

the same time, even if supplements are given at different times than the main feed, it does little good because the animal body cannot store amino acids not suitably balanced for protein production even for a few hours."[11]

But how are we to know what amino acids are actually provided in the feeds we use? Lack of this or that one can severely inhibit growth, reproductivity, the development of healthy tissue as opposed to fat, and on and on. Unfortunately, without a broad base in nutrition, food science or agricultural science and the guidelines set by the National Research Council, we can't. There is nothing to indicate to us on the labels of the feeds either the percentage of the amino acids available and usuable, nor that required by the National Reserach Council. The only thing we laymen can know is that a certain percentage of "all the essential elements of a canine diet" must be contained in the manufacturer's "recommended amount of feed per day;" however, the latitude in the percentage of each requirement may vary enough—sometimes as much as 10%—that the balance is upset and the resulting "protein level" can drop drastically because it cannot be utilized by the animal's body.

Soy bean products are highly prized by human vegetarians, particularly for the protein. As a result, soy products are a major component of many kibble feeds. But they can cause severe problems in a canine because they can interfere with the use of iodine by the thyroid gland, signs of which can be seen in poor coat, anemia and sterility. They, along with beets, can also produce canine bloat (a problem which is seen to be increasing) because of the creation of gas in the intestines caused by fermentation of these materials. A carnivore's natural foods in the past, meat and bone, do not produce saponins, which seem to be responsible for this problem.

If you look at the component tables in Appendix D, you will find that most of the manufacturers are relying on grains as the source of protein, fiber and starches. The cheaper feeds depend to a great extent on the gluten contained in corn. The more expensive feeds utilize a greater ration of soy bean derivatives, beets, and in some few, meat by-products are listed as primary ingredients.

What then is the reasoning behind the use of grain rather than meat as the primary diet for a carnivore with teeth made for tearing and chewing and an appendix and gut structure built for digesting hunks of meat that are not chewed?

One very obvious reason is that grains are much cheaper and can be cooked, compacted and marketed in less volume and at much less expense than meat. One widely touted justification is that "in the wild, wolves will first eat their prey's stomach and entrails," thus ingesting the undigested grains the prey has eaten—therefore it must be more desirable and necessary to the predator. This philosophy totally disregards the opposing ideology that the stomach is the softest, most vulnerable and most accessible part of the prey when it is downed and thus would be eaten first. It also disregards the fact that very little of the carcass is not eaten and that the order of eating probably has little or no importance in judging what is necessary to the predator's diet. Do not humans seem to prefer junk foods and sweets, though they are not nutritionally in need of them? They're fast, easy and very accessible. This bit of reverse anthropomorphism is not meant to imply that the the stomachs and entrails are "junk foods"—simply that the easy accessibility once an animal is down doesn't seem to have been taken into account in the manufacturers' logic.

Even kibble feeds that do use meat products as their principal ingredient have their detriments. Any processed food, whether dry, semi-moist or canned, loses a good bit of its nutrient value in the manufacturing process. Just as in the foods we eat, the more it is cooked, the less vitamin and nutrient value survives. The ingredients of dry kibble would, of course, suffer the most, since the ingredients are cooked, dried, compacted—a series of processes which reduces the nutrient values in each process. Thus, as in processed human foods, artificial vitamins must be added to replace those that have been depleted through processing, (those indecipherable chemicals listed on the package labels), along with artificial odors and tastes. Even a diet of pure cooked meat loses some because of the cooking process.

Scientists have told us that perhaps the most perfect balance of protein and nutrients exists in a raw, whole egg. Obviously you can't feed your animals tons of raw eggs a day, and such a diet wouldn't provide the bulk and fiber which the digestive system also needs, but the addition of raw eggs to the diet you use would certainly improve it. This is one reason for the widely heard "give your animals raw eggs to improve their coat." Again, lack of sufficient and balanced protein produces many symptoms, among which are a noticeably poor, dull coat.

Too little protein, or imbalanced, unusable protein in the diet can produce many problems and symptoms, perhaps the harshest being that of infertility. Animals deficient in protein may exhibit poor, dull coats, skin problems, lack of muscle development, loose stools or diarrhea, infertility, stunted growth and increased susceptibility to disease. Conversely, an over-abundance of protein may result in the animal being unable to use the protein and simply passing it in a loose stool, or by becoming obese from "overfeeding."

There is another factor to consider in evaluating the protein (or other component) level of any given dog food. That is the "recommended amount" to be fed per pound per day. The more bulk that is required to provide the amount of usable nutrients, the lower the nutrient value. Thus, if for one particular feed you must feed 8 cups of kibble to a 70 pound dog to achieve a 21% protein level, and for another only five cups, the nutrient level is obviously higher in the second, and while it may be more expensive, where is the savings over the course of a week if you have to feed twice as much of the first to achieve the same result?

Additionally, the dog may not be able to or interested in consuming that much bulk and where it becomes primarily bulk, the majority of it will simply be passed without too much absorption by the body.

As far as protein is concerned, where lies the perfect answer? Obviously, in a raw meat, fowl or fish diet. Unfortunately, this is not something available to most owners. Not everybody lives in comparatively remote areas where "road kills" are often found, or near a slaughter house where very cheap trimmings are available. Not all of us can afford freezers for our animals' needs alone. And there is the fact that our canines seem to be able to do well on various diets that feed manufacturers have produced for them.

THE ECONOMICS OF NUTRITION
(Reprinted from *The Wolf Hybrid Times*, April 1988)

If the various kibbles do provide "adequate maintenance diets" and if a natural diet provides a better balance for a wolf or Wolf Hybrid, what are the economics of each? How much of either does a wolf, a Wolf Hybrid or a dog of similar weight and stress need for maintenance and for optimum health?

We are advised that on the Yukon Quest endurance race, racers are required to carry 2 pounds of feed per day per dog. The Steger Expedition to the North Pole carried the Science Diet Maximum Stress Formula and provided 1½ pounds per day per dog. Some of these animals were Hybrids, some Canadian Huskies. These working dogs are under tremendous physical stress and obviously the diets they are fed are going to be high protein.

Where quantities of feed are measured and fed according to the manufacturer's directions, an animal weighing between 75 to 100 pounds would do well on 6 to 8 cups of kibble per day (depending on the particular brand of kibble). Quantitatively, for animals on a pure meat diet, we are advised that approximately 2½ lbs. per day with some chicken neck "snacks" provides very excellent results.

The effectiveness of various diets can be judged by many factors:

1. The health and general "tone" of the animal on any given diet;

2. The ability to utilize a particular type of diet may often be reflected by the animal's personality—much like a human—i.e., a person who decides to go on a "crash diet" may become very irritable simply because of deprivation of essential nutrients;

3. The amount of feed actually utilized by the animal, i.e., the amount of unusuable waste expelled by the animal represents money spent for quantity, a great deal of which will be cleaned up from the pens, not food that the animal's body can assimilate and use. On the other hand, an animal which is *gradually* changed to a highly usable diet will excrete less, develop better tone and function in stressful situations—stressful either physically or psychologically.

Several subscribers have written asking the age old question—"Won't an animal that has tasted blood or raw meat become more aggressive?" This theory has been applied to animals of all species through the generations. There is no evidence to support such a theory. The taste and smell of raw meat to a carnivore will always be more appealing than a synthetic version. Isn't steak more appealing to you than hamburger helper?

Humans eat meat. Some prefer to be vegetarians because of intellectual or emotional preference. There is no evidence that meat-eating humans are more aggressive than vegetarians. Likewise, there is no evidence to show that either man or other meat-eating animals are more aggressive if they prefer their meat rare (or raw) rather than well done.

Quite naturally, a "chicken on the run" is far more inviting than a dead one or one that has been slaughtered—for a different reason. The movements of the "chicken on the run," a mouse or a rabbit tearing across the field stimulate a natural prey instinct. Again using reverse anthropomorphism, long time hunters may shoot a standing deer, elk or what have you, but the real excitement and sense of accomplishment comes from the chase and felling an animal on the run. Repugnant? But true. The senses are excited by the flight, the challenge. True, these hunters may eat the meat after proper treatment of the carcass, but many don't. Given a well-fed animal, the carnivore is similarly excited by the chase.

In the years of raising many breeds of dogs and Wolf Hybrids, we have never encountered an overall change of personality when the diet consisted of raw meat as opposed to manufactured feeds. The animals who were super possessive of their food expressed the same behavior, whether over raw meat or kibble. Whatever the feed, it was theirs and they would defend it against people or any other animals. When the food was gone, the normal behavior remained unaffected.

Based on our survey, animals fed on a "free feed" basis (once adapted to it) seem to consume less than those fed a rationed amount daily and are less uptight about it (i.e., especially at feeding time). This applies equally to those animals on raw meat diets and those strictly on kibble diets. Many subscribers have written from areas where they have access to "road kills" or donations from farmers or ranchers of animals that have been killed or died, and simply leave pieces of the carcasses in the pens until consumed (sometimes in conjunction with free-feed kibble) who report actual daily consumption being less and body utilization of the foods greater (i.e., reduced excrement, better body tone and reduced excitement).

To go back to economics—most of the various commercially produced feeds will maintain your animal nutritionally, consumed in the quantities recommended on the packages. These amounts are based on the percentage of the food that the animal's body will utilize—not on what will fit into its stomach. If one feeds kibble mixed with a canned food plus a supplement (costs not contemplated here) and most of the bulk of that diet is expelled and not utilized by the body, are you really saving any money by feeding the large quantity of an artificial diet as opposed to a pure meat diet which can be fed in less quantity with much more utilization.

This is a question which only you as an owner can answer. You are the only one who can on a daily basis determine if 1) this or that diet is serving your animal better (remember to give it at least a 30-day trial), and 2) if it is economically feasible for you to give the very best. Check out the packing houses and meat markets near you for trimmings. Nowadays, not many throw them away, but most are cheap. Very few owners can afford to feed on supermarket prices, but chances are, even with costs of shipping, you'll come out even or pounds ahead by buying frozen packaged meat "by products" from the packing houses that offer it.

The following charts are based on actual weight of cup measures and the manufacturer's recommendations for feeding animals of 75 to 100 pounds, as effective in April, 1988.

KIBBLE:

Based on the manufacturers' recommendations for feeding a 75# to 100# animal:

Dealers Pride (Purina)—5 cups per day. Sold in 50# sacks at approximately $15.00. (5 cups weigh 1.25 lbs.)

Iams Eukanuba—6 cups per day. Sold in 40# sacks at approximately $25.00 (6 cups weigh 1.625 lbs.)

Science Diet—6 cups per day. Sold in 40# sacks at $28.96 per sack. (6 cups weigh 1.625 lbs.)

	Cost Per Day	Cost Per Month	Cost Per Year
One Animal:			
Dealers Pride	$ .38	$ 11.40	$ 136.80
Iams	1.02	30.47	365.63
Science Diet	1.18	35.30	423.54
Five Animals:			
Dealers Pride	1.90	57.00	684.00
Iams	5.10	153.00	1,836.00
Science Diet	5.90	177.00	2,124.00

If you add one can per animal of canned food, add approximately 50¢ per day per animal, $15.00 per month or $178.00 per year per animal. If you also feed vitamin supplements, you must also add that cost to your figures.

PURE MEAT DIET—feeding 2½# per day per animal.

at .20/lb.	.50	15.00	180.00
at .30/lb.	.75	22.50	270.00
at .50/lb.	1.25	37.50	450.00

PRICES OF PURE MEAT DIETS FROM VARIOUS BY-PRODUCTS PACKERS AND SLAUGHTERHOUSES

Hereford By-Products
Hereford, Texas
Beef By-Products at 26¢/lb., shipped in 5# blocks.

Nebraska Brand
A feline diet at 48¢/lb., shipped in 5# blocks.

Champion Quality Meats
Alaska
Beef By-Products at 35¢/lb.
Other meats range from 35¢/lb. to 67¢/lb.

The above figures are very simplistic and are only minimum bases of comparison. We all tend to provide extra to our animals and perhaps much more than they need or can utilize. Costs of supplements are not taken into consideration in any of the above figures, nor is a plus-or-minus factor figured in.

It is a simple task to weigh out how much of any diet you are using, figure the cost per pound and compare it to the charts. A 50 lb. sack costing $15.00 would be figured:

$15.00 divided by 50 = 30 cents/lb.

A 50 lb. sack costing $21.00 would be figured:

$21.00 divided by 50 = 42 cents/lb.

You may be very surprised at the results. Given the tremendous difference in costs in most instances, and the degree of unutilized and eliminated food, some owners may find that the savings in one year from feeding a pure meat diet from one of the slaughterhouses or by-products distributors could well buy a sizable freezer plus greatly enhance the health and well-being of their animals.

CHAPTER XVII. CONTAINMENT

FROM INTERNATIONAL CITY MANAGEMENT ASSOCIATION "REPORT," October, 1976
Volume 8, Number 10

"The control of animals, especially dogs and cats, in an urban environment is one of the most widespread problems facing municipalities throughout the world. These pets originally served a crucial economic and community health purpose by controlling the population of rats and mice, and indeed are still important aids in that battle, but the conditions of living in cities and towns have made these protectors as big a nuisance as the pests they help control."

"Pets, especially dogs, now represent a major public health threat themselves. Over 1.5 million dog bite incidents are reported to authorities every year. Free-ranging animals scatter trash from containers, causing litter, and providing breeding grounds for flies, roaches and rats. Animal feces and urine create a major sanitation problem."

The laws in almost all urban localities require *adequate* containment or restraint for any pet. No matter what type of fencing or containment facilities are used, if the animal can get out, it is obviously not adequate.

When one is dealing with a Wolf Hybrid, one must keep in mind the potential size and strength of the animal, as well as a sometimes very fierce determination. A three to four foot fence may seem adequate for a puppy, but then again, I have seen some determined pups climb six foot fences with ease.

One sees all manner of fencing for pets—some that is so ridiculous that it makes one shudder. Goat fencing, horse fencing, "garden fencing"seem to some to be all that they would need...after all, their pets don't *want* to run away. In reality, this type of fencing is so pliable that the wires can easily be pushed aside to make a very neat escape hatch. One good rule of thumb is that if YOU can do it, so can THEY! Fencing designed for sheep, goats or cattle would be just like so many cobwebs to a Hybrid, let alone a pure wolf.

Many people rely on high wooden fencing, some with great success, but if you are dealing with an escape artist, this too has its drawbacks. As most are aware, Hybrids love to chew on wood. Witness the trees in most Hybrid owners' yards, ringed with fence! Any small hole in the fence will most likely get larger. Also, if the support rails are on the inside, they are just dandy "ladders," for kids and dogs alike.

A dandy ladder for kids and dogs alike. Photo: D. Prendergast.

At Rudelhaus, we use tautly stretched six foot high heavy gauge chain link fencing with an additional 1-½ feet of smooth (not barbed) "tilt wire" on top, like the security fencing commonly seen around commercial business yards. This is tilted inward to the animal compounds. I emphasize "tautly stretched" because chain link that is even slightly loose can be as easily bent upward or outward as weaker type fencing.

In addition to this vertical fencing, four feet of fence is laid on the ground inside of the yards and tied tightly to the chain link and then covered with dirt, rock and other heavy materials to prevent digging. Owners commonly refer to this as "ground wire."

When my fencing was first installed, I thought that would probably do the trick. Wrong! I had a jumper in the house. She showed me very handily and quickly that it still wasn't enough. It was then that we installed a single electric wire about one foot in from the fence just at nose level. It only took two stings by the wire to make her realize that she didn't want anything further to do with the fence, no matter *what* was stinging her. Some people add a hot wire at the top of the fence. Keep in mind, though, that in order for the electric wire to do its work, whatever touches it must be grounded in order to get stung. If the animal has its feet on the fence when it touches the wire, then it will get stung, but if it is leaping through the air and touches the wire, nothing will happen at all.

Where nothing else works for a determined climber or jumper, perhaps the only alternative is the very expensive type of enclosure recommended for zoos or exhibits—with a solid roof, combined with chain link tied tightly into the side fencing over the entire enclosure.

An equally important consideration is the problem of deterring digging. An inherent trait by itself, when directed at tunneling out of the yard, the problem becomes paramount. While some are able to afford deep, concrete footings around their yards, others use wire mesh ground wire with rock or other more creative ''ground lining.'' If it works—use it. The old line ''it ain't much for pretty but hell for stout'' might make some of the more beauty conscious shudder, but where the security of your animal is concerned, it's worth considering.

It goes without saying that simply erecting a stout and sturdy fence will never be the end of it. Fences must be continually checked for weakness or loosening, and reinforced, added to or whatever is necessary to make sure that the animals are kept inside. Lax maintenance can be disastrous because just one weakness can mean the escape and possible loss of the animal.

The following article appeared in *The Wolf Hybrid Times* in April, 1987, and is partially based on information supplied by Mary McBee, a long-time breeder well versed in the building and moving of pens.

''An enclosure or facility does not have to be beautiful by human standards to be adequate and to satisfy the animal's needs. There are many real 'junkyards' which do fulfill this function very well, but which many of us couldn't live with in our backyards. Beautifully landscaped areas do much to satisfy the owners and prospective buyers, but an animal really has no preferences as to beauty.

Many levels in the enclosure do much to enhance the beauty of an enclosure and wolves and Wolf Hybrids do appreciate this sort of variety. Most dogs, as well as wolves and Hybrids, love to play 'king of the mountain,' and generally prefer the most elevated spot to lay and sun or sleep, but this can be accomplished as easily with sturdy tables, the flat top of their 'dog house' or den, a sturdy wooden box of adequate size, or a huge spindle (the type used for transporting heavy wire cable that many of us convert to picnic tables).

Water is necessary, and while some live in areas where a stream can be diverted to run through the pen giving both fresh drinking water and a pool, most of us do not have this advantage. Many people provide drinking water by simply making sure fresh water in buckets is always available or by a drip water fountain connected either directly to an outdoor faucet or a large, suspended 'bottle,' and provide 'fun spots' by setting up either a child's wading pool or stock tank for cavorting in the water. The latter is preferable because many times the claws penetrate the rubber or plastic pool causing it to leak. Worse, it is more likely to be torn apart and ingested with disastrous physical results.

Perhaps the most important part of containment is psychological. One animal may be very content to live in an outdoor 25' x 50' enclosure if, when its owner is home, it also lives indoors and in close proximity. This may also depend on the percentage of the Hybrid. For another, this would be totally unacceptable and it may quickly develop the personality of a zoo animal, pacing constantly, howling (even barking), being fairly destructive and always looking for a means of escape.

Escape can be a learned behavior. As Mary McBee put it:

'If a pup is placed in a good, securely fenced compound

Two year old male, percentage not reported. Rocky Mountain Timber cross. Photo: Jacci N. Bayard.

when young, it rarely 'learns' to try escaping. The Hybrids (or wolves or dogs) that are 'escape artists,' I've found, are usually ones who were in poor and inadequate fencing when young and got out rather easily—so they were more or less 'taught' that if they looked hard enough, there would always be a way out. But at times, owners unwittingly 'teach' their animals to be escape artists and to climb. For example, put a pup in a 3' fence. Then when he learns to climb that, put it up to 5'; when he learns to climb that, put it up to 7', etc. Put it up at 8' to begin with while it's small.''

What type of fencing material should be used? Light weight (or lighter gauge) 2" x 4" welded mesh is a waste of money. The welds practically fall apart when you unroll it and animals walk right through it—even poodles. But again, one has to make decisions: what you can afford, and for what size animal. Perhaps the most important consideration is the height and ''gauge'' of fencing used. The lower the number of the gauge, the stouter the fence. You must also look at the size of the opening between the wires. Is it too wide? Can the younger animals climb right through it? Is the wire merely welded, twisted, or is it welded at each twist besides? Is it steel or is it hog wire. Many use lesser fencing at great risk—particularly during mating season.''

HOT WIRE

Electrically charged fence wire is used by many, sometimes without understanding of what they are actually doing. It will definitely discourage attempts to come near the wire—the first couple of jolts will teach them that this is something to respect. But many people do not realize the intensity of the impact of the ''jolt'' unless they experience it themselves. There are different types of chargers for electric fences. Some require a 6 volt battery, some a 12 volt battery and others are designed to be hooked up directly to household current.

A 12 volt battery charger will put out a shock that will burn the skin of a human, and not only leave him tingling, but numb for several minutes. Think what it would do to your animals, especially if their coats are damp. Most charges are designed to transmit an electric current for a minimum of a half mile, with much longer minimum ranges being available.

If you have a fairly small area and choose to use the electric wire even intermittently, after a few "bites" by the wire, the animals will not willingly "challenge" the wire again, but most likely will stay 4' to 6' from it. How much room do they actually have left? Are they going to be left actually inhabiting an area that most of us consider unhealthy and undesirable for a wolf or Wolf Hybrid? We all hate putting our dogs in a kennel to board because of the limited space they are confined in. It's important to consider whether, by cutting down on their actual living area, we have not created our own kennel.

CONTAINMENT IDEAS FOR THE SAKE OF THE OWNER

In today's mobile society, it seems rare that a family remains in the same house for its entire life. In fact, it is more common for families to move more than five times during their lives together, often many times more. Fencing is expensive and physically draining to erect, especially as you get older. Whether you are renting or own your own property, fencing becomes "part of the property" when it is permanently sunk into the ground. Many people need to modify their enclosures, either to accommodate additional animals or simply to change the arrangement initially built. This necessarily results in a lot of money and work, both to tear down existing structures and to build new ones. There is a cheaper, and in the long run, less strenuous way to go about it.

The concept of "modular" homes is not new, but the conversion of this idea to fencing may be quite new to some. Although it may be initially more expensive to build "panels" of fencing, these panels may be moved, transported, converted to use in enclosures of any dimension or shape with the only additional expense being that of sinking the fence posts to which they are attached and transporting them when they need to be moved. The design is simple and can be modified to suit what sort of transportation you may have available. The dimensions suggested here are based on transporting a number of panels standing vertically in a standard pickup truck.

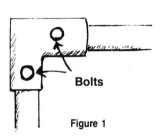

Bolts

Figure 1

Initially, the "framework" must be constructed. This can be accomplished by cutting and welding heavy rods or chain link-type fence posts together into either 6' x 10' or 8' x 10' rectangles. Or instead of welding them, one may purchase the standard elbow fittings, drill holes in both sides of each "corner" and bolt them securely.

The panels should not be larger than this if they are to be transported in the back of a standard pickup standing vertically, and larger panels would also be extremely difficult to move because of their weight. The use of wood for the framework of the panels is not recommended for two reasons: 1) the weight of such materials and 2) their vulnerability to destruction by weather, insects or chewing.

The next step is stretching the wire. This is something that cannot easily be accomplished by hand. "Fence stretchers" can be rented from most tool companies, lumber yards or catalogue sales companies. It is important that the fencing be stretched tight; otherwise, they are flimsy, easy to bend, easy to tear and in some cases, the chain link twists around and becomes entangled. Some people accomplish the trick by using heavy clamps and turnbuckles and/or their vehicle to put enough tension on the fencing to pull it tight.

The fencing must be clamped securely to the framework. One can use the commercially made stretcher rods or concrete rebar and clamps on the vertical sides, and tying each "joint" to the framework at the top and bottom. Whichever the choice, it must be remembered that the edges of the panels are the most likely areas to be tested for escape and are also the most vulnerable because of stress to the fencing material, so it is important that care be given to the even attachment of the "fabric" to the framework.

Once the panels are constructed the fence posts can be installed. Remember to include in your measurements the dimensions of the fence post and the space on both sides of the fence posts which will be consumed by the clamps or bolts holding the panels to the pots. The fence posts should be much heavier and of a wider dimension than those used in the construction of the panels—they may have to take a lot of stress.

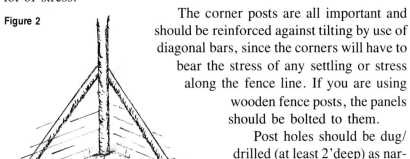

Figure 2

The corner posts are all important and should be reinforced against tilting by use of diagonal bars, since the corners will have to bear the stress of any settling or stress along the fence line. If you are using wooden fence posts, the panels should be bolted to them.

Post holes should be dug/drilled (at least 2'deep) as narrowly as possible and filled with concrete around the posts.

Corner post holes should be sunk much deeper and should be greatly reinforced by a "collar" to ensure their strength and reliability. It is advisable to allow the concrete to "set" for several days before attaching the panels.

The line and corner posts have been set, your panels are constructed and the concrete has aged for a couple of days. Now you are ready to stand your panels. The panels can be attached to the posts by any secure means, while still making sure that they can be detached when desired. If you have used standard chain link posts, all that is needed is to use the commercially available clamps and bolts to attach the panels to the posts. When you move or want to change the arrangement of your panels, you can simply remove the clamps and rearrange the panels as you wish.

Following this, skirting or ground wire should be laid upon or buried into the ground, but firmly attached to the vertical fencing to prevent digging. Tilt or security fencing can be attached to the top of the panels to prevent climbing or jumping, and for determined escape artists, the top of the enclosure may be covered with fencing.

GATES.

Gates can be built into a panel with no problem. It is highly recommended, particularly if there is concern about a particular animal or animals getting out or getting together, to use a double gate system whereby there is a 3' x 6' enclosed and covered "entryway" to the area with a secure gate on each end, each with drop latches. See Figure 3. In the event an animal should accidentally slip into the entryway, it is prevented from going any further by the second gate. This is particularly helpful during breeding season when you need to control the interaction of the animals (and people).

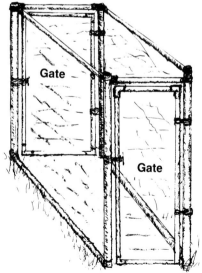

Figure 3

Same construction as described,
but this panel is framed
out of wood.

In the end, fencing/containment comes down to providing for the health, security, safety and social well being of both animals and people. An animal should never be kept on a chain. It could never be capable of developing the type of personality we are aiming for, and an animal that escapes finds itself in immediate danger from humans and/or the mechanical wonders of our modern civilization.

PANELS AND CLAMPS

A circus came to town the other day, and while the kids concentrated on the performances and the antics of the clowns, I found myself concentrating on the workers who would put up, and then take down the cages for the lions, tigers, giraffe and other animals in the circus. While watching what was actually a very speedy process, I was immediately reminded of Mary McBee's suggestions on building fencing in panels.

The huge ''ring'' for the cats was constructed of 5' wide x 12' high panels made of very heavy gauge 4'' square welded wire which, in turn, was welded onto fairly light-weight (but apparently steel) frames. The frames had cross-bars running at two foot intervals.

The ''clamps'' which held the panels together were what intrigued me the most. they were about four feet long with hook-like ends which fitted around the cross-bars of each frame, and had a clamping device, much like the handles of the old refrigerators, to lock them tight. I discovered that the insides of the panels and wires had been painted black, possibly for the benefit of the animals—so that they could see them against the bright lights and know their limitations.

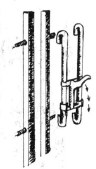

Pull lever down to clamp sleeve around cross-members of the two panels.

Although these panels were not locked into the ground in any way, my mind could visualize the security such panels would provide were they securely attached, and visualized the same construction, only 8' x 16' standing horizontally and tied into permanent corner posts. Next time the circus comes to town, check it out!

XVIII. DIGGING

DIGGING—WHAT CAN YOU DO?

Reprinted from *The Wolf Hybrid Times*
June 1987

WARNING

This article is intended partially in fun. The photos used in this article are courtesy of wildlife photographer Monty Sloan and were all taken at the Folsom City Zoo in California. Obviously, they are (at least partially) man-made dens and not your common everyday Wolf Hybrid type "lark digging."

The photos feature 1) Lupine and Terra, 2) Lupine, 3) Terra('s tail), 4) Lupine and Terra.

Lupine and Terra, Folsom City Zoo. Photo: Monty Sloan.

Lupine, Folsom City Zoo. Photo: Monty Sloan.

I have been in hundreds of other "wolf yards" before. I KNOW what to expect. But each time I go visiting someone else's yards, the same thing happens. I am so intent on devouring the animals and their environment with my eyes that I invariably end up ankle (or sometimes knee) deep in a hole and spending the next few days disgustedly hobbling to and from the refrigerator for more ice cubes, trying to bring the swelling down.

You'd think that after some thirteen years I'd learn. I don't have that trouble in my yards—but probably because here, I'm conditioned to look DOWN instead of up!—except for keeping a watchful eye for sneaky tackles or clipping maneuvers from behind.

It is while I'm confined to the couch or the overstuffed chair that I remember all those letters I chuckle at and which almost beg for a reason why their animals dig, hoping to stop them.

Aside from broken legs, people who have written to us have expressed several major concerns. A lot full of holes is just not cosmetically beautiful, and where Hybrid owners are renting the property, it can be very upsetting to the landlord.

Another worry heard mostly from people in the southwest is how to protect the trees. To someone living in the east where greenery is often something you dig up to make room for a garden, a house or whatever, this may not seem like much of a problem, but in the western states where getting anything to grow is a 24 hour a day labor of love, it is a major concern. If a Hybrid opens a large air space, even without damaging the roots, the tree is quite likely to die because of the roots drying out. This probably means not only the loss of a sizable investment of time and money but also the loss of natural shade for the animal who contributed to its demise. Also in relation to trees and bushes, Hybrids seem absolutely locked in on the idea of digging up and ingesting any fertilizer which may be applied below the surface—a very dangerous activity.

The simplest solution to the above is to fence in the trees. You may say that this will not look any too nice. On the other hand, I have seen some pretty imaginative and beautiful "tree guards," even in yards where there are no Hybrids or other large animals—beautiful stained or painted picket fences, circular or square symmetrical enclosures high enough to prevent chewing of the bark and far enough the trunk to prevent digging in under.

Yard digging has to be expected. All dogs do it occasionally—sometimes out of boredom, sometimes just for the lark, sometimes chasing a rodent, real or imagined.

Terra's Tail, Folsom City Zoo. Photo: Monty Sloan.

Lupine and Terra, Folsom City Zoo. Photo: Monty Sloan.

A concern often voiced is over denning and worrying that the tunnels might extend beyond the fencing and perhaps surfacing on the outside of their enclosure. Unfortunately, tunnels to the dens do sometimes extend over ten feet through the ground. We haven't heard from anyone whose animals created a second entrance outside the fence, but we have been told of several incidents where, since the animals had little choice of terrain or type of soil in which to dig their dens, the dens have collapsed after a rain, killing infant cubs inside. Many people elaborate on the dens the animals start, and put wooden roofs over them covered by dirt, leaving the rest as was. This would certainly give the owner more confidence in the dens.

The largest concern, of course, is over the ''digger'' who is intent on getting out of its enclosure. The questionnaires returned so far indicate that while most people do use ground wire, they also feel that if the animals are content and not bored in their enclosures, they do not try to dig out. Almost all indicated having companion animals in the enclosures to help maintain this lack of interest in escape.

One innovative response, reminiscent of Farley Mowat's *Never Cry Wolf*,[12] suggested burying feces in any hole that was started, covering it with dirt and then urinating it as 90% effective for them.

There is no one answer to curbing the instinct to dig, and with the exception of the use of "ground wire," perhaps better not to try, so long as the digging is not an effort to escape.

CHAPTER XIX.
PREVENTATIVE MEDICINE

An Article by Ken Podkonjak, D.V.M.

Reprinted from *The Wolf Hybrid Times*
February 1986

VACCINATIONS.

Wild and domesticated and hybrid canids are susceptible to all the clinical problems of the domestic dog. The management of these problems is likewise quite similar.

In the Canidae family, of which there are 37 species, including the coyote, wolf, fox and dogs, there are several infectious diseases of clinical importance to the Hybrid owner. They include canine distemper, canine corona virus, canine parvo, canine hepatitis, rabies, leptospirosis, toxoplasmosis and tracheobronchitis or parainfluenza. In my clinical experience, the Wolf Hybrid seems especially sensitive to viral diseases. While I cannot specify why this seems to be the case, I can relate that in my practice the Husky and Malamute breeds appear to be more sensitive to anaesthesia, infection and parasitic diseases. One may surmise that the Wolf Hybrid may have acquired this sensitivity through hybrid crossing with these breeds, or perhaps the pure wolf, because of original lack of exposure, is yet unadapted to "modern society's" diseases. Nevertheless, the important message to susceptibility is prevention. Often by the time the owner recognizes sickness in an animal, the disease process is far advanced and can be incurable. This is especially true with distemper and parvo disease. I cannot over-emphasize the importance of an adequate prevention plan.

I like to approach a disease prevention plan by considering all factors that may influence or retard the development of a disease. This includes:

1. Exposure
2. Adequate vaccination programs
3. Stress prevention.

An often overlooked consideration in disease prevention is avoiding exposure during the critical, unprotected time in the pup's life. This is as critical as an adequate vaccination program. The Hybrid owner must realize that one vaccination at six to eight weeks of age does not immediately protect the pup. In addition, even with a series of vaccinations, the pup may not be fully competent to fight diseases until twenty weeks of age. This helps emphasize the importance of not exposing the pup to potentially sick or convalescent dogs. One must avoid bringing new dogs into the household or kennel during this time. The owner should also avoid visiting other kennels, dog pounds or privately owned dogs with questionable vaccination and disease-free status. Realize that the germs can be carried by you on your hands, clothing and shoes to your pup.

In questioning a great number of my clients whose puppy has contracted distemper or parvo disease, there commonly has been

Kodiak—reported 94% Eastern Timberwolf/Alsatian male. Photo: Kitty & Gerry Whitehouse.

exposure to a stray pup brought into the household. Perhaps the neighbor's dog was sick and the pup came into contact with it (or its owners), or the pup went to a facility where animals congregate, like animal control centers, boarding kennels or dog care facilities. Try to plan the arrival of other animals to your facilities at times other than whelping time. If this is not feasible, do the best job you can in quarantining any new arrival completely away from your other animals for thirty days.

One of the biggest mistakes I experience is people purchasing animals with inadequate vaccinations. Not only do you jeopardize the new animal investment (which is often substantial), but you may jeopardize your entire investment in Wolf Hybrids. Try not to purchase new animals which have an inadequate vaccination history.

This brings us to the second phase of our "prevention plan," adequate vaccination programs. An always controversial subject is animal vaccine. Which types are best, when is the best time to vaccinate and what should we vaccinate against, are common questions.

In considering a vaccination program, I don't think there is any one program that has proven to be far superior to another. Researchers speak of antibody titer levels, protection even in the presence of maternal antibody levels from the milk and all sorts of fancy scientific terms, often directed to a particular vaccine type. I believe the shortcoming in many of these "claims" is failure to apply research to particular areas, especially problem areas such as particular animal types, i.e, Wolf Hybrids, Malamutes, Huskies, Dobermans, Blue Heelers, Rottweilers, who seem particularly sensitive; and a well-timed, yet economical vaccination schedule. One must realize that a program for Beagles in a research colony may not be applicable for Wolf Hybrids in an urban environment. The points I would emphasize are:

1. Be sure the whelping bitch is current on distemper, parvo, hepatitis, leptospirosis, coronavirus, parinfluenza, rabies and current to within six months of whelping.
2. Schedule pups to begin vaccinations no later than five weeks of age.
3. Insure boosters are carried out beyond fourteen to sixteen weeks of age. A regime that has worked well for me in an area plagued with distemper, corona and parvo is to vaccinate pups with half-doses at five and seven weeks, and to boost every two weeks

until fourteen to sixteen weeks of age. This regime is keyed toward high exposure risk. The problem with this is expense to the owner, so the program is rarely utilized, even though it is safer.

4. Selection of vaccine best suited for your individual situation. It is important to note that KILLED rabies virus vaccines only are to be used on any wolf or Wolf Hybrid. Rabies breaks have been experienced in modified-live rabies vaccines (MLV). Since the wolf (and its hybrids) are not approved for rabies vaccine use in any event, don't use the MLV.

Currently, there is a lot of controversy concerning the type of vaccine to use. I will not endorse brand names; use what is working best for you in your area. I will say that I have been involved in vaccine problems in my area. In my experience, strictly modified live vaccines (MLV) have given me some serious problems in sensitive breeds, including the Wolf Hybrid. I have had far greater success using products that combine modified live and killed canine and feline cell line origin vaccines. I have also backed away from the use of distemper-measles (MLV) in pups because of notable problems. I prefer now to use the same vaccine for all age groups.

A common mistake I find is the client philosophy that one shot provides instant protection that lasts forever. Remember, the shots do not become effective for ten to fourteen days, and pup boosters and annual vaccination are essential to prevent disease.

STRESS PREVENTION

The third aspect of our "prevention plan" is stress prevention. Pups sometimes undergo tremendous environmental and internal stresses. Weaning is a stressful condition. To add traveling, dietary changes, internal and external parasites, environmental temperature and humidity changes, vaccinations and children playing with the pup non-stop, may be pushing our luck a bit. The point is to avoid unusual amounts of stress, to plan changes in an orderly fashion with rest periods, and avoid drastic changes in temperature, food and care. This may help reduce susceptibility to disease problems.

PARASITES

External and internal parasites should both be considered in parasite control programs. Common external parasites include fleas, ticks,

Shadow—reported 75% wolf. Photo: Mollie Peterson.

mites and lice. External parasites can represent a bothersome, chronic problem or a moderate to severe acute problem. The larger bugs (can be seen with the naked eye) are usually controlled with prescription preparations that can be topically applied. Fleas, ticks and lice are visible to the naked eye and do respond quite well to approved insecticidal powders sprinkled sparingly on the entire body at weekly intervals. In my experience, powders, used properly, seem more effective than flea collars, especially on longer haired, larger animals. (Not to mention the problem of your other Wolf Hybrids chewing the collars off). To detect larger parasites, I recommend regular grooming, looking for "small, dark colored, moving creatures."

Mites are not visible to the naked eye. They are commonly referred to as scabies and mange. Diagnosis of skin mites are commonly made by clinical observation coupled with microscopic examination of skin scrapings. Because mites tend to live in the skin, they are much more difficult to control and treat. Experimentally, there are injectable drugs that are quite effective in treating mange. I have had good success with preliminary trials using injectable anti-ectoparasitics.

Any time you notice excessive scratching, hair loss or reddened skin, I recommend a visit to your veterinarian. In addition, consider sterilizing the ground and surfaces of your run areas.

Internal parasites, especially in the puppy, can be a significant health problem. Rather than worrying about specific worm types, the owner should consider routine fecal examination and deworming schedules. It is best to check with your local veterinarian for specific parasite problems indigenous to the area in which you live, like hookworm and heartworm. Round worms and tapeworms are very common in pups and you should always include parasite control in your health program.

We have discussed the importance of an adequate vaccination program, avoiding excessive stresses like drastic changes in environment and feeding, disease exposure and parasite control. Although I feel that these, along with an adequate diet, are the most important aspects to consider as a guide to Hybrid health, they are merely guidelines. Each Hybrid is an individual that may require special attention. Utilize your observations and those of professional as well as experienced lay people to learn all you can about your Hybrid. With caution and care, you can better enjoy and respect a truly beautiful animal.

CHAPTER XX. EMERGENCY CARE AND FIRST AID

Often the most difficult decision for owners when they see something unusual in their animals is whether it warrants veterinary attention. If there is a question, be safe and see a veterinarian at once. The following first aid tips are intended only as *temporary* measures until the animal can be gotten to the veterinarian.

ABDOMINAL DISTENSION

Abdominal swelling can be caused by many things, some being more severe than others. Simple over-eating in pups is very common and usually is not serious. Gastic, canine bloat or stomach twist and dilation are absolute emergencies and are almost always fatal without immediate veterinary attention.

Over-eating and over-drinking can occur at any age but are common in young pups, or in animals that have not eaten regularly. This can be caused by unaccustomed, sudden consumption of large quantities of food, eating garbage, consuming excessive amounts of water after ingestion of large amounts of dry dog food or following exercise or unavailability of water for an abnormally long time.

Signs: Swollen abdomen; may vomit or wretch; groaning and uncomfortable behavior.

Treatment: Stop all food and water for approximately twelve hours *BUT* if distension is severe or respiratory distress is present, see your veterinarian at once. Stomach twist and dilation are most often seen in large, deep chested animals. Shock, coma and death can occur within two hours. The cause seems to be related to ingestion of a large meal in some instances but not in others. Signs in these cases are more severe with 1) enlarged abdomen; 2) extremely painful abdomen

113

that may sound like a drum when thumped with the forefinger; 3) excessive salivation and unsuccessful vomit attempts; 4) difficult breathing; 5) shock; 6) reluctance to move, refusal to lie down until collapsing. Once down, often will not move. THESE ARE VETERINARY EMERGENCIES! DO NOT WAIT—GET IMMEDIATE HELP.

BLEEDING

Perhaps bleeding causes panic more than any other emergency. Remember, animals sense panic. Remain calm and use proper, safe restraint to avoid the animal's over-excitement and excessive bleeding.

Arterial bleeding can represent critical blood loss. It is spurting in time with the heartbeat and bright red. Veinous bleeding tends to ooze and is darker colored; however, in excess, it can represent danger. Severe bleeding must receive immediate attention.

Direct pressure or pressure bandaging should be tried first to control bleeding. Tourniquets are LAST RESORTS ONLY.

Direct pressure over a wound using a clean cloth is a very effective hemmorhage control technique. Minor cuts will usually stop within a few minutes. On serious cuts, direct pressure should be maintained during transport to a veterinary facility. Sanitary napkins or disposable diapers are excellent for this purpose. They are clean, provide padding, thereby helping to avoid wraps being too tight, and are readily available. Four inch strips of cloth or three inch gauze in a roll can be used to hold the padding in place. It is best to pull just firmly with each wrap (not too tight), then tape over the entire wrap with any adhesive tape available.

With severely mangled or crushed areas that cannot be managed by direct pressure, a tourniquet can be applied. Two inch gauze or cloth can be applied above the wound but taking care not to tie it too tightly. RELEASE EVERY TEN MINUTES and get to a veterinarian as soon as possible.

Bleeding ears are common. These are best managed by bandaging the ear down to the head. Food pads are commonly cut and bleed a lot. Diapers work great! Put the pad in the center and pull the diaper up on all sides of the leg. Wrap firmly, including the foot, and seek veterinary attention.

Management of internal bleeding is done primarily at the veterinary clinic, but it is important to treat for shock. Keep the animal quiet, warm, free of mucous, blood or vomit in the air passages and transport immediately. Pale to white gum color is a good indication of shock and possible internal hemmorhage, along with the obvious signs of coughing up blood or blood in the excrement. Do not give any water or food.

BURNS

Burns are occasionally a problem. Superficial burns are usually characterized by singed hair, skin redness and mild swelling. Treatment includes preventing licking and scratching of the area, cold water compresses for fifteen minutes, gently removing overlying hair and applying topical antibiotic ointment.

Deep or major burns should receive immediate veterinary treatment. Try to keep the animal as quiet as possible. Prevent licking and scratching of burn areas. Bandage the involved area with clean, dry gauze and treat for shock if indicated.

Shadow—reported 89% British Columbian cross. Photo: Gene & Pat Guenther.

115

CHOKING

Choking is caused by obstruction, throat swelling and allergy-type reactions. Kennel cough is not choke, although many people think it is. Signs of choking are sudden difficulty in breathing, cyanasis (blue to purple tongue and lips) and collapse.

Wind pipe obstruction is an emergency and the dog should be placed on its side and the rear portion of the rib cage struck firmly with the palm of the hand (not so severely as to injure the animal, however). Repeat if necessary. It helps to pull the tongue forward. Artificial respiration may be necessary.

Obstructions and foreign bodies in the mouth and esophagus (food pipe), are common. The animal often looks apprehensive, salivates, paws its mouth or rubs its head along the ground. While some obstructions may be removed by the owner, veterinary assistance is recommended.

CONVULSIONS

Convulsions (seizures) are violent, involuntary contractions of muscles. There are many causes and the severity of convulsions may vary. Signs include restlessness, a dazed look, twitching, salivation and lip licking, dilated pupils, loss of consciousness and violent muscle contractions. Frequently, urine and bowel control is lost.

Continuous seizures require immediate medical attention. Owners must NOT try to physically restrain the animal and children should be kept away from it. Never place a finger or hand between the teeth. The animal will not "swallow its tongue." Do not attempt to give anything to drink. Try to blanket and pad the animal and move any sharp objects, etc., which may be injurious to it.

All seizure cases should be referred to a veterinarian.

EAR INJURIES

Ear injuries are a common problem. Bites, wire cuts, foreign bodies in the ear canal, parasitic and bacterial ear canal infections are frequently seen by veterinarians.

Signs of inner ear problems include violent head shaking, scratching at the ears and neck, dragging ears along the floor, head tilt, discharge, swelling and tenderness.

Ear cuts can result in a lot of blood loss because of head shaking. A pressure wrap around the head, including the ear, will help the hemmorhage problems. Any ear canal problem should be seen by a veterinarian as soon as possible. Do not probe the ear with Q-Tips, etc. A mild, gentle mineral oil flush may help relieve pain temporarily until veterinary help can be obtained.

A monthly cleaning with a mild oil will help prevent excess wax and dirt build-up as well as infection problems.

EYE INJURIES

Eye injuries should be immediately treated by a veterinarian. Foreign objects can occasionally be flushed out with Visine or other eye wash, but do not attempt to pull foreign bodies from the eye, as injury to the cornea is possible.

Chemical irritants can be flushed with water for ten to fifteen minutes—then seek veterinary attention at once.

In the case of a prolapsed (disengaged) eyeball, both eyes should be covered with cold compresses to help control eye movement and animal taken to the veterinarian. The eyeball should be replaced within an hour.

FRACTURES, DISLOCATIONS, SPRAINS

By definition, *closed fractures* are those in which the skin has not been penetrated by bone. *Open fractures* have punctured or torn skin associated with bone ends. *Dislocations* are displacement of one or more bones associated with a joint. *Sprains* are tearing or stretching of soft tissue (i.e., ligaments) around a joint.

In general, one or more signs will be associated with these conditions, including: inability to use the limb, either partial or complete; pain around the injury site; swelling and bruising; grating (a sensation of bone edges rubbing together; bone sticking through the skin.

The owner should immediately: prevent further injury and infection by proper transporting techniques; should *not* attempt heroic first aid practices, such as trying to reset fractures or use of elaborate splints that 90% of the time are improperly applied; should protect himself with the use of mouth ties or muzzles; and should simply wrap the animal in a blanket to control shock and movement and transport immediately to a veterinarian. Control hemmorhaging if necessary, using a clean wrap over the wounds.

117

PUNCTURE WOUNDS

Most puncture wounds are incurred from bites of other dogs or cats, or by stepping on sharp objects. If the animal has stepped on a thorn or a fox tail has become imbedded in the foot or skin, simple removal of the thorn or fox tail may be all that is required. The area should be carefully monitored, however, to be sure that no abscesses develop, which will have to be treated by a veterinarian.

If the animal has stepped on a nail or other sharp object, it should be removed at once, and administration of antibiotics begun to prevent infection.

If the animal has been bitten, antibiotics should also be administered and the bite locations carefully monitored for infection or abscess. In all puncture wounds, it is important that healing take place from inside the wound before it is allowed to seal over. Should swelling occur, veterinary treatment should be sought immediately.

Annastasheena and friend, percentage unreported. Photo: Joan Fleetman.

FOOTNOTES

1. The Hour of the Wolf—that short span of time between first light and dawn when the spirits roam the earth; when the wolves howl most.
2. Miglionardi, Mario, *German Shepherds,* Arco Publishing Co., N.Y., 1971.
3. Mech, L. David, *The Wolf, the Ecology and Behavior of an Endangered Species,* University of Minnesota Press, 1970.
4. Hall, E.R. and K.R. Kelson, *The Mammals of North America,* Vol. II, The Ronald Press, NY, 1959.
5. Lopez, Barry Holstein, *Of Wolves and Men,* Scribner, 1978.
6. Frank, Harry, *Man and Wolf,* Dr. W. Junk Publishers, Dordrecht, The Netherlands, 1987.
7. The Wildlife Education and Research Foundation (WERF), a non-profit corporation, Gallup, New Mexico.
8. Klinghammer, Erich, *Wolf Literature References,* Wolf Park, Battle Ground, IN, 1989.
9. Wolf Park, Battle Ground, IN.
10. Mission Wolf, Silver Cliff, CO.
11. *Feeding For Results, a New and Critical Approach,* Robert Abady, Robert Abady Food Company.
12. Mowat, Farley, *Never Cry Wolf,* Little, Brown, Boston MA, 1963.

APPENDIX A

VARIOUS WOLF HYBRID ORGANIZATIONS AND REGISTRIES IN THE UNITED STATES

Please note that this listing is not all inclusive and that mailing addresses are current only as of January, 1989.

Alaska Wolf Pack
Box 671741
Chugiak, AK 99967

Aniwaya Wolf Club of Texas
PO Box 233
DeSoto, TX 75115

A.W.H.A.
Box 217
Larchwood, IA 51241

California Wolf Hybrid Club
5785 Little Uvas Rd.
Morgan Hill, CA 95037

Iowolfers Association
Rt. 4, Box 215-A
Mount Pleasant, IA 52641

N.A.H.W.A.
SR Box 5197
Wasilla, AK 99687

Northern California Wolf Pack
1436 La Vista Avenue
Concord, CA 94521

Northwest Wolfers
2923 Gretchen Way
Boise, ID 83704

Universal Kennel Club
P.O. Box 574
Nanuet, NY 10954-0574

U.S. Wolf Hybrid Association
Box 351-A, HC-87
Delhi, NY 13753

APPENDIX B
GENETICS

The following two articles are reprints from the April, 1988 issue of *The Wolf Hybrid Times* and were written in response to an article printed in *Dog World* in January, 1988, in which some rather erroneous information regarding genetics and Wolf Hybrids was quoted. The information provided by the authors may give a broader understanding of genetic probabilities without being unduly weighty.

GENETICS IN REAL LIFE

by Jerold S. Bell, D.V.M.

Reprinted from *The Wolf Hybrid Times*
April, 1988

The following is a critique of an article, "Ambassador Wolves" which was published in the January 1988 issue of *Dog World*.

The article in general was well written and included mostly factual data. However, there were a number of misleading statements concerning the genetics of dog-wolf hybrids. The figure on page 54 is reportedly a graphic representation of 10 of 78 genes. In actuality, it is a representation of ten *chromosomes,* or five chromosome pairs. There are over 100,000 genes which are carried on 78 chromosomes in the wolf and domestic dog (39 chromosome pairs).

The concept of genetic variability involves the dual process of: 1) the pairing of genes from two genetic parents to produce an offspring, and 2) the recombination of genes on chromosomes through the mechanism that takes a 100% sire and 100% dam and makes a 50/50 offspring. The less understood mechanism of genetic crossover is responsible for trading the genes on the chromosomes, so that genes for traits carried on a single chromosome are not always inherited together.

During the first stage of meiosis (production of the egg or sperm cells) the single strands of the chromatids containing the gene pairs may overlap and trade sections. Simply speaking, this means that when the egg or sperm are produced, the *combinations* of genes on one chromatid may vary between the individual and the egg or sperm produced. Note that this crossover occurs *before* fertilization and combination with the genes from the opposite mate. Therefore, genetic crossover is not a mechanism to receive both genes in a pair from one parent. An individual homozygout (having identical genes in a gene pair, i.e., "AA") for a particular gene can only pass on an "A" gene to all of its offspring and will never (barring a very rare mutation) pass on an "a" or produce an "aa" individual.

Due to genetic crossovers, the chromatid in the egg or sperm of the wolf-dog hybrid will never contain purely wolf or dog genes as represented by a black or white circle in the figure on page 54 (of the *Dog World* article). All ten of the circles should realistically be colored grey. As the genetic material of the dog and wolf are now intimately mixed on the chromosomes, the chances or probability of getting a 100% wolf from a wolf x wolf-dog hybrid mating, or a pure dog or pure wolf from a mating between two wolf-dog hybrids is infinitely small. For the same reason, the statement in the fifth paragraph on page 52 stating that a mating between a German Shepherd and a wolf-dog hybrid may produce a "pure dog anyway" is realistically impossible.

Statistically speaking, the product of a wolf x wolf-dog hybrid mating will produce an offspring with approximately 75% wolf and 25% dog genes. Genetic crossovers in the hybrid parent will blend the dog and wolf genes together on the chromatid, but remember that the crossovers occur *before* fertilization mixes the genes of the two parents. For each crossover that adds more wolf genes to the sperm or egg, the probability is the same for a crossover on another chromatid to switch dog for wolf genes. The probability of *significant* varying from the standard 75/25 ratio is extremely small, and the probability of segregating all of the wolf genes into one offspring is realistically impossible.

The notion of getting a "genetic throwback" to a purebred animal from a hybrid cross occurs because of the *appearance (phenotype)* of hybrids resembling and possibly even matching the behavioral characteristics of a purebred dog or wolf. This is due to either the dominance of certain genes over others, or the segregation to homologous genes (identical genes in the pair) *of that particular gene or group of genes.* While the genes responsible for the appearance and/or behavior that is being observed may be more like the purebred dog or wolf, genes for other observable or non-observable characteristics (enzymes, blood types) will be of the opposite type in the hybrid.

Lastly, while the product of hybrid crosses may resemble a purebred animal, it will not reproduce those characteristics cleanly, due to its hybrid genetic makeup.

* * * * * * * * * *

THE HOWLING

Peter P. Mullen, M.S., MT (ASCP)
Professor of Laboratory Science

Reprinted from *The Wolf Hybrid Times*
April 1988

PREFACE:

Like a lonely wolf overlooking an unknown territory, I am howling loudly in hopes of communicating with any intelligent life lurking in my domain. I am howling about the erroneous statements concerning canine genetics appearing in the January 1988 issue of *Dog World.*

I am howling about the incredibility of the article entitled "Ambassador Wolves"...

I am howling about the quote, "the reason we're here is to educate the public on the true nature of the wolf"...

I am howling about the confusion of genes and chromosomes.

I am howling about the lack of knowledge concerning phenotype and genotype.

I am howling about the scientific inaccuracy concerning wolf-dog crosses.

I am howling loudest about the statement—"Furthermore, when you mate two 50% wolves you may wind up with anything from a pure dog to a pure wolf."

In order to appreciate the concepts of canine genetics, it is necessary to begin with the basic study of animal cells. An animal is entirely made up of cells consisting of two distinct parts—the nucleus and the cytoplasm. The cytoplasm is responsible for cellular metabolism while the nucleus is primarily concerned with body growth and cell division.

Within the nucleus of a canid cell are 39 pairs of long rodlike structures known as chromosomes; these pairs of chromosomes are said to be homologous, meaning identical in terms of structure. The numbers of chromosomes in any one species is constant, thus identifying it as a biological species at the molecular cellular level. In canids (dogs, wolves and coyotes, etc.) the full complement of chromosomes consists of 78 chromosomes or 39 pairs.

Chromosomes are composed of long strands of deoxyribonucleic acid (DNA) on a protein framework. The basic physical structure of DNA is seen as two long polynucleotide chains forming a double spiral. These chains are composed of nucleotides, each nucleotide being composed of one pentose sugar, one phosphate and one organic base.

The chromosomes within the cell's nucleus are the bearers of hereditary factors; these units of heredity are called genes which are arranged in linear order on the chromosomes and each gene occupies a specific position on the DNA chain known as a locus. Segments of the DNA molecules therefore compose the genes which are coded for three base sequences known as codons. Each codon codes for a particular amino acid. A series of codons determine the amino acid sequence on a polypeptide chain and a gene is a sequence of codons that contains the code for one polypeptide.

A particular locus may be occupied by several different forms of the gene; these alternatives are known as alleles. When these allelic genes on a pair of homologous chromosomes are identical, the canid has inherited a gene coding for the same characteristic from both parents; this is referred to as a homozygous trait. When the alleles are different, the canid is said to be heterozygous.

Genes carry from generation to generation the information that specifies the characteristics of a particular species. The product of a gene is the characteristic it produces. The blue of a canid's eyes can be considered the product of the particular gene determining that characteristic.

A canid obtains its chromosomes and genes from each parental sex cell called a gamete. Each gamete donates 39 single chromosomes (not genes) thus giving the fertilized zygote a full complement of chromosomes totaling 78. Each offspring is composed of 50% of the mother's genetic material and 50% of the father's genetic material. The one-cell zygote duplicates itself by mitosis and passes the genetic information on to all the progeny cells. The embryological development of the canid cub is influenced by the DNA molecules which have randomly combined making each sibling a unique individual. The number of chromosomes (78) is what is responsible for controlling the zygote to develop into a canid rather than a feline. The number of genes does not influence this embryological process.

The processes of reproduction are governed by two separate cell alterations known as meiosis and mitosis. The first, meiosis, is concerned with the production of sperm or ovum, the gametes. The second, mitosis, is the process of cell division by which the canid grows and replaces discarded cells.

The sex cells are responsible for the initiation of reproduction of the entire organism. The production of sperm and ovum is known as gametogenesis, a process that results in the production of a cell possessing only half the number of chromosomes present in a normal body cell. Some references call this reduction division. If nature did not govern meiosis in this manner, each offspring would receive a full complement of chromosomes from each parent, thus doubling the chromosome number for each generation. The chromosome number for a given species would never be constant.

CONCLUSION:

In conclusion, it must be emphasized that chromosomes are responsible for passing on genetic material to offspring, not genes. The genes are a segment of the chromosome that influence individual traits. The chromosomes are the transporting structures of the DNA molecules. The number of chromosomes, not the number of genes, identifies the species. Because each offspring is composed of 50% of the mother's chromosomes and 50% of the father's, the zygote is composed of exactly one-half of each parent. The physical traits, called phenotype, may resemble one parent more than the other, but the genetic traits, called the genotype, are an equal combination of both.

References:

Bryant, Neville J., *An Introduction To Immunohematology*, W.B. Saunders Company, 1982.

Dog World, MacLean Hunter Publication, January 1988, Vol. 73, No. 1, Page 16.

APPENDIX C

BOTTLE FEEDING

by Dorothy Harpster

Reprinted from *The Wolf Hybrid Times*
June, 1988

There has always been considerable controversy on whether or not to bottle feed wolf and Wolf Hybrid pups. Of course, occasionally there are medical reasons that necessitate this. If the mother has mastitis, a mammary gland abscess or hypocalcemia (too little calcium), then the pups must be removed and bottle fed at all too often an early age for their own nutritional requirements.

The controversy stems from many experts feeling that removing the pups at an early age and bottle feeding socializes them better to humans. Erich Klinghammer[1] states that wolf pups are best removed from their mother well before they are three weeks old. By the time wolf pups are 21 days old, they are so fearful that socialization to humans is very difficult. He prefers taking them at about ten days of age.

Jim Rieder[2] of the Timberwolf Preservation Society prefers to take pups at twelve days of age to bottle feed. And there are certainly many other wolf experts, including Erik Zimen, Harry Frank and Ulysses Seal that concur.[3]

Dottie Prendergast[4] clearly states that she is a naturalist in her views on this matter. And who can argue when she says that a natural weaning creates a healthier pup? What follows is not intended to argue for or against bottle feeding; rather it is an attempt to show the best nutritional way.

It has been well documented that young wolf pups fed on canine milk replacer exclusively may develop cataracts.[5] Jim Rieder and his wolves were involved in a five year study to determine the effects of supplemental milk formulas. The Department of Opthalmology at the University of Illinois was asked to assist the wolf breeder since many of his pups fed solely on this commercial milk replacement were blinded with cataracts. It was decided that using Esbilac[6] caused cataract formations during the critical time of lens development. These lens opacities may develop as early as seven days after the pup is placed on the formula. The biochemistry of nutritional cataracts is not completely understood. However, the analysis of Esbilac, dog and wolf milk (Figure 1) will show you that wolf milk is higher in percentage of all amino acids.

It is immediately obvious how nutritionally deficient canine milk replacer is for wolf pups. Most wolf and Wolf Hybrid breeders are well aware that cow's milk is an extremely poor substitute—after all, Mother Nature intended it for baby cows, not baby wolves. But many breeders over the years have felt that Esbilac alone was a nutritionally complete diet for nursing wolf and Wolf Hybrid pups.

Esbilac is extremely low in amino acids compared to natural wolf milk. An amino acid is a building block of protein. Therefore, it is necessary to supplement canine milk replacer to avoid cataracts and possibly other deficiencies. The findings suggest that diets deficient in arginine result in cataract formation in young timber wolves.[5]

Wolf pups and Wolf Hybrid pups should not be removed from the mother before ten days of age unless there is some legitimate medical reason. Some say ten days, some say twelve days. Perhaps waiting until their eyes have opened is a good alternative. Hopefully, the lens will be fully developed by then. The earlier the pups are started on the milk replacer, the more severe the lens opacities that have been observed. This suggests there is a critical time at which the lens is more susceptible to nutritional stress.

There are several variations recommended by the experts. You could use one part Esbilac powder to three parts water with Knox unflavored gelatin added to provide arginine to prevent the cataracts. Another version is again one part Esbilac powder to three parts water and 15 grams or 15% lactose, dry form. Others suggest supplementing even more with the addition of one part half-and-half and with a teaspoon or two of honey or Karo syrup. They feel additional fats and

Figure 1

AMINO ACIDS (mg/gm)

	Esbilac	Wolf's Milk	Dog's Milk
Lysine	2.81	5.49	4.32
Histidine	0.937	3.55	3.08
Arginine	1.24	7.344	5.51
Aspartic Acid	3.04	11.20	7.86
Threonine	1.82	4.84	4.08
Serine	2.21	5.50	4.39
Glutamic Acid	8.94	27.30	21.90
Proline	4.15	13.60	11.40
Glycine	0.696	0.922	0.795
Alanine	1.28	3.92	4.01
Valine	2.89	7.03	5.36
Methionine	0.326	2.65	2.40
Isoleucine	1.95	5.79	4.81
Leucine	3.92	17.10	14.00
Tyrodine	2.02	4.56	3.75
Phenylalanine	1.89	5.64	4.47
Tryptophan	0.969	2.01	2.40

AND

	Esbilac	Wolf's Milk	Dog's Milk
Galactose	0.30	0.12	0.21
Lactose	2.60	1.80	2.46
Glucose	0.01	ND	ND
Ash	0.83	1.65	1.20
Phosphorus	0.110	0.276	0.190
Potassium	0.148	0.178	0.109
Sodium	0.071	0.062	0.088
Calcium	0.149	0.418	0.278
Manganese	0.010	0.018	0.010
Protein	5.1	12.60	10.70
Ammonia	0.694	2.56	2.03

ND = None Detected

carbohydrates are needed, especially during growth spurts in pups at three and four weeks when blood volume increases very rapidly. Many breeders as well as veterinary opthalmologists report normal lens development in pups reared on milk replacer supplemented with either fructose or glucose (honey or Karo syrup). Both honey and Karo syrup have 60 calories per tablespoon and honey has 17.3 grams of carbohydrate to Karo syrup's 16 grams.

When artificially nursing, use standard rubber nipples and bottles for human babies. The size of the nipple's hole is important. The pups might aspirate (inhale into their lungs) the formula if the hole is too large. They could also stop nursing before ingesting enough if the hole is too small. It may well be advisable to wear a glove of some type on the hand holding the bottle as they will knead you with their front paws, causing numerous scratches. If you are feeding six pups every four hours, your hand could become rather sore.

At twenty-five days of age, solid food can be introduced. Feeding a moistened dry puppy kibble to which is added a finely ground canned puppy food is nutritionally sound. The addition of either water or milk replacer with thorough mixing to create a gruel consistency, should be carried out until five to six weeks of age.

It should be noted that the above table was printed in 1981. I have before me two cans of Esbilac. The liquid (ready to feed) states protein at 4.5%, whereas in the powdered form, the protein content following reconstitution is lower (4.2%). The liquid is certainly more expensive at $4.00 per 12 oz. The powdered form is $12.00 (12 oz.) for an amount equal to 96 oz. liquid. It is also available in a larger size can for less money per ounce. The amount of protein has certainly not increased over the years. At the present time, there is nothing that duplicates nature's finest—wolf or Wolf Hybrid's own milk.

In *Man and Wolf,* the book recently reviewed in this publication, Harry Frank makes a dedication. He says, after naming three wolves:

"Who taught me to understand CANIS (LUPUS).

"One must first learn to see the world through golden eyes."

This leads me to think that one needs to understand the nutritional requirements of baby wolves and Wolf Hybrids so that they may indeed see through golden eyes.

References:

1. Klinghammer, E. (Editor) 1979. *The Behavior And Ecology Of Wolves,* Garland, STPM, New York.
2. Reider, J., 1985, *Wolf News,* Vol. 5, No. 2.
3. Frank, H. (Editor) 1987. *Man and Wolf,* Dr. W. Junk, publishers, Dordrecht, The Netherlands.
4. Prendergast, D., 1984, *The Wolf Hybrid,* Rudelhaus Enterprises, Gallup, N.M.
5. Vainisi et al, 1981, "Nutritional Cataracts In Timber Wolves," *Journal of the American Veterinary Medical Association,* Vol. 197, No. 11.
6. Esbilac—liquid—Borden, and powdered manufactured for Pet Ag, Inc., 201 Keyes Ave., Hampshire IL 60140.

APPENDIX D
COMMERCIAL FOOD INGREDIENTS

The following are actual ingredient label claims as they appear on the packages and are reproduced so that the reader may make comparisons and draw his own conclusions regarding the content of each feed.

ABADY BLUE RIBBON
(High Stress & Competition Formula)

Crude Fiber—not less than 25.0%
Crude Fat—not less than 19.0%
Crude Fiber—not more than 0.5%
Moisture—not more than 8.5%

Ingredients: Poultry by-product meal, Menhaden fish meal, kibbled yellow corn, kibbled wheat, ground toasted corn, ground toasted wheat, soybean oil, meat & bone meal, lard, rice, wheat bran, monobasic calcium phosphate, beef fat, corn starch, dessicated liver, sodium chloride, fish protein concentrate, potassium chloride, Torula yeast, egg albumen, whole dried egg, choline chloride, d-Alpha tocopherol acetate, zinc oxide, ascorbic acid, vitamin A acetate, inositol, niacinamide, d-Calcium pantothenate, para-aminobenzoic acid, cobalt chloride, cholecalciferol (source of vitamin D_3), cupric oxide, pyridoxine hydrochloride, thiamine hydrochloride, menadione (source of vitamin K), sodium selenite, folic acid, biotin, cyanocobalamin concentrate (source of vitamin B^{12}).

BENCH AND FIELD 26

Crude Protein—not less than 26.0%
Crude Fat—not less than 9.0%
Crude Fiber—not less than 3.0%
Moisture, not more than 11.0%

Ingredients: Meat meal, kibbled corn, cereal food fines, kibbled wheat, corn gluten meal, ground extruded whole soybeans, dehulled soybean meal, dried beef pulp, corn distillers' dried grains with solubles, bentonite, animal fat (ethoxyquin preservative), condensed corn fermentation solubles, dried whey products, dried A oryzae fermentation extract, vitamin A supplement, D-activated animal sterol (source of vitamin D_3), vitamin E supplement, riboflavin, calcium, pantothenate, niacin, choline chloride, vitamin B_{12} supplement, menadione sodium bisulfite complex (source of vitamin K), thiamine mononitrate, sodium selenite, calcium iodate, zinc oxide, manganese sulfate, copper oxide, cobalt carbonate, iron, sulfate, methionine hydroxy analog calcium.

PURINA O.N.E.

Guaranteed Analysis:
Crude Protein—not less than 26.0%
Crude Fat—not less than 16.0%
Crude Fiber—*not more than* 3.0%
Moisture—not more than 12.0%

Ingredients: Chicken, ground yellow corn, ground wheat, corn gluten meal, poultry by-product meal, beef tallow preserved with ethoxyquin, brewers' ground rice, dried beef digest, dicalcium phosphate, salt, choline chloride, calcium carbonate, L-lysine, wheat germ meal, zinc oxide, ferrous sulfate, niacin, vitamin supplements (A, D_3, E, B_{12}), calcium pantohenate, ethoxyquin (a preservative), vegetable oil, manganous oxide, riboflavin supplement, biotin, folic acid, copper oxide, thiamin mononitrate, pyridoxine, hydrochloride, garlic oil, inositol, menadione sodium bisulfite (source of vitamin K activity), calcium iodate, cobalt carbonate.

IAMS EUKANUBA

Guaranteed Analysis:
Crude Protein—not less than 30.0%
Crude Fat—not less than 20.0%
Crude Fiber—not more than 4.0%
Moisture—not more than 10.0%

Ingredients: Poultry by-product meal, ground corn, chicken, poultry fat (preserved w/BHA) beet pulp, meat meal, rice, brewers dried yeast, dried whole egg, monosodium phosphate, potassium chloride, DL-methionine, salt (sodium chloride),choline chloride, ascorbic acid (vitamin C), vitamin E supplement, copper sulfate, zinc oxide, manganese sulfate, manganese oxide, biotin supplement, vitamin A acetate, vitamin B_{12} supplement, calcium pantothenate, riboflavin supplement, thiamin mononitrate, niacin, menadione dimenthylprymidinol bisulfite (source of Vitamin K), copper oxide, inositol, pyridoxine hydrochloride (vitamin B_6), vitamin D_3 supplement, potassium iodide, folic acid, cobalt carbonate, sodium selenite.

PURINA DEALERS PRIDE

Crude Protein—not less than 21.0%
Crude Fat—not less than 7.0%
Crude Fiber—not more than 6.0
Moisture—not more than 12.0%

Ingredients: Ground grain, sorghum, ground yellow corn, soybean meal, ground wheat, wheat middlings, meat and bone meal, animal fat preserved with BHA, salt, calcium carbonate, dried whey, wheat germ meal, zinc oxide, ferrous sulfate, brewers dried yeast, vitamin supplements (A, D_3, E, B_{12}), DL-methionine, manganous oxide, niacin, garlic oil, pyridoxine hydrochloride, copper oxide, thiamin mononitrate, folic acid, menadione sodium bisulfite (source of vitamin K activity), calcium iodate, cobalt carbonate.

133

SCIENCE DIET
MAXIMUM STRESS DIET

Crude Protein—Min. 27.0%
Crude Fat—Min. 25.0%
Crude Fiber—Max. 5.0%
Moisture—Max. 10.0%

Ingredients: Animal Fat (preserved with BHA, propyl gallate, citric acid), soy flour, meat meal, dextrose, dried whole egg, dried skimmed milk, vegetable oil, brewers dried yeast, dicalcium phosphate, calcium carbonate, choline chloride, iron sulfate, zinc oxide, manganous oxide, copper oxide, cobalt carbonate, vitamin A palmitate, D-activated animal sterol, a-Tocopherol, niacin, calcium panthothenate, riboflavin, thiamine, pyridoxine hydrochloride, folic acid, biotin, vitamin B_{12} supplement, potassium sorbate, polyoxyethylene sorbitan monooleate.

SCIENCE DIET
CANINE MAINTENANCE

Crude Protein—Min. 22.0%
Crude Fat—Min. 13.0%
Crude Fiber—Max. 4.3%
Moisturee—Max. 10.0%

Ingredients: Ground corn, meat meal, soy grits, brewers rice, animal fat (preserved with BHA, propyl gallate, citric acid), wheat bran, dried whole egg, brewers dried yeast, vegetable oil, iodized salt, calcium carbonate, iron sulfate, zinc oxide, manganous oxide, copper oxide, cobalt carbonate, choline chloride, vitamin A, palmitate D-activated animal sterol, a-Tocopherol, niacin, calcium pantothenate, riboflavin, thiamine, pyridoxine hydrochloride, folic acid, biotin, vitamin B_{12} supplement.

WAYNES

Ingredients: Meat and bone meal, ground corn, grain, wheat, brewers rice, corn gluten meal, animal fat (preserved with BHA), soybean meal, wheat middlings, L-lysine, salt, brewers dried yeast, dried whey product, D-activated animal sterol (source of vitamin D_3), vitamin B_{12} supplement, vitamin E supplement, menodione sodium bisulfite complex (source of vitamin K activity), riboflavin supplement, niacin supplement, biotin, calcium pantothenatee, choline chloride, thiamine, pyricosine hydrochloride, folic acid, cobalt carbonate, iron carbonate, copper oxide, manganous oxide, sodium selenite and preserved with BHA, propylene glycol, propylgallate and citric acid.

CANNED FOOD COMPARISONS

ALPO BEEF & LIVER DINNER

Guaranteed Analysis:
Crude Protein—not less than 11.0%
Crude Fat—not less than 5.0%
Crude Fiber—maximum 1.0%
Moisture—maximum 78.0%

Ingredients: Meat by-products, water sufficient for processing, beef, liver, poultry by-products, soy flour, salt, caramel color, potassium chloride, guar gum, methionine hydroxy analogue calcium, DL-alpha tocopheryl acetate (source of Vitamin E), citric acid and ethoxyquin (preservatives) magnesium oxide, choline chloride, iron carbonate, copper oxide, cobalt carbonate, manganous oxide, zinc oxide, ethylenediamine dihydroiodide, thiamine mononitrate, D-activated animal sterol (source of Vitamin D_3 and Vitamin B_{12} supplement.

KAL KAN Pedigree Chopped Combo with Chicken, Beef & Liver

Guaranteed Analysis:
Crude Protein—Min. 8.0%
Crude Fat—Min. 6.0%
Crude Fiber—Max. 1.5%
Moisture—Max. 78%

Ingredients: Meat by-products, water sufficient for processing, chicken, beef liver, poultry by-products, fish, sodium tripolypyphosphate, vegetable gums, salt, potassium chloride, whole egg, calcium sulfate, artificial color, caramel coloring, artificial and natural flavors, onion powder, garlic powder, zinc sulfate, Vitamin A, D_3 and E supplements, tetrapotassium pyrophosphate, ethoxyquin (a preservative), calcium pantothenate, thiamine mononitrate (B_1).

CANNED HUSKY DOG FOOD

(Note: this canned food makes no claims of being a meat product.)
Crude Protein—minimum 8.0%
Crude Fat—minimum 2.0%
Crude Fiber—maximum 2.0%
Moisture—maximum 78.0%

Ingredients: Soybean meal, ground corn, poultry, beef by-products, cracked barley, soya hulls, wheat bran, iodized salt, iron oxide, vitamin A supplement, vitamin D_3 supplement, vitamin E supplement, cobalt chloride, potassium iodide, D-calcium pantothenate, riboflavin supplement, thiamine mononitrate, vitamin B_{12} supplement, biotin, seasoning.

APPENDIX E

SYMPTOMS OF SERIOUS DISEASE

It cannot be emphasized too strongly that the life of your Wolf Hybrid depends on the *prevention* of disease. A regular and complete immunization and worming schedule is absolutely essential to protect it from the diseases described below which can kill an animal within hours of the first noticeable symptoms.

CANINE DISTEMPER

Since the discovery in 1923 that canine distemper is caused by a virus, research has provided us with an effective means of preventing this fatal and common disease, provided the immunizations are received on a timely and regular basis.

The virus of distemper affects all tissues and organs of the infected dog. Each animal may react differently to the infection. In many cases, the affected dogs show symptoms like a "cold." There may be a watery discharge from the eyes and nose, coughing, lack of interest in food, and the dog may have a fever. Many develop pneumonia. One of the most serious kinds of distemper occurs when the virus attacks the central nervous system. Symptoms may not develop until 30-40 days after the dog was exposed, but then convulsions or seizures may occur and perhaps muscular twitchings known as chorea. These neurally stimulated signs indicate a very serious condition and the dogs rarely recover.

INFECTIOUS CANINE HEPATITIS

This disease is caused by an adenovirus and is not related to human hepatitis. The symptoms are usually drowsiness, lack of appetite, high temperature and abdominal tenderness. There may also be a cough and tenderness in the throat region; however, it is not unusual for an unprotected dog, particularly puppies, to die of this disease without any recognizable symptoms.

LEPTOSPIROSIS

The symptoms of this disease are very variable and may be mild or severe. In severe cases, there may be a high fever, discharge from the eyes and nose, muscular soreness, a yellow to dark orange discoloration of the eyes and skin, with vomiting and possibly death. It is possible, although rarely, for man to be infected from dogs carrying this organism. The disease is passed through direct ingestion of the virus from food and water dishes or from humans who have handled sick dogs. It is also passed in the urine of infected dogs and can be so transmitted for several months after the dog has recovered.

PARVOVIRUS

One of the most severe and prevalent canine diseases is parvovirus. Since it was first recognized in 1978, its incidence has skyrocketed and is most often fatal once the symptoms become apparent if immediate treatment is not initiated, particularly in young pups. The incubation period for parvovirus is three to twelve days.

The fatality rate in young puppies is high, but can be reduced in older puppies and adult dogs if treatment with supplemental fluids is undertaken immediately upon recognition of the symptoms. These include sluggishness, high fever, vomiting and diarrhea, both of which can be persistent and contain blood. The result is usually rapid dehydration and death.

The virus is spread primarily through contact with the feces of an infected dog. Antibiotics have no effect on viruses. Parvovirus is resistant to most detergents and alcohol, and can withstand freezing and high temperatures.

CORONAVIRUS

The symptoms of this disease are very similar to those of parvo and without "high tech" detection equipment, one disease may be confused with the other. Corona tends to be even more deadly than parvo; however, immediate treatment is generally the same. Again, prevention through vaccination is by far the best means of saving the animal.

RABIES

Rabies is probably the most well-known and feared of any canine disease. It is transmitted from infected animals or carriers through a bite or the transfer of saliva containing the virus.

The symptoms are variable and are usually marked by noticeable changes in behavior. These changes may range from a very sluggish or uninterested attitude in a normally very active to hyperactivity in a normally calmer animal. The animal abruptly loses its appetite and seeks to be alone. The early symptoms may involve more frequent urination and discomfort with swallowing and thus the reluctance or inability to swallow saliva. Subsequent changes in one to three days may show signs of viciousness or paralysis.

Most communities now require an animal to have current rabies vaccinations because of the threat to humans. There is, as yet, still no vaccine on the market which is specifically licensed as providing protection for wolves and Wolf Hybrids. The research which has been conducted for the last two years by The Wildlife Education Research Foundation (WERF), as previously mentioned, indicates that the vaccines on the market do indeed provide the necessary protection, and I would urge *all* wolf and Wolf Hybrid owners to use these vaccines. It cannot be too strongly emphasized that the rabies innoculation is as necessary as the DHLP, Parvo and Corona innoculations and that if a killed vaccine is used, it will not bring on a case of rabies unless the animal already has rabies to begin with. It is absolutely essential that *all* the immunizations are kept current and the animal be protected against these very deadly diseases.

IMMUNIZATION SCHEDULE

It is essential that the Hybrid be immunized against all of the above diseases on a regular schedule. Breeders normally provide the "puppy shots" that provide temporary protection against distemper, leptospirosis and measles, but as indicated in Chapter XIX, some veterinarians are no longer recommending the measles shots. They are leaning more and more toward giving half doses of the regular vaccines for the first couple of shots and then beginning the full annual doses. These shots must be kept up on a regular schedule if they are to be effective.

The puppy should begin receiving half doses of DHLP and parvo at five weeks with a repeat half dose every two weeks until they are at least nine weeks old. At Rudelhaus, we begin the full dosage at nine weeks and continue to provide this protection every two weeks until they are at least fourteen weeks of age. Boosters for all the diseases can be given annually thereafter.

The first rabies innoculation should be given at three and one-half months of age, with periodic boosters thereafter, as indicated by your veterinarian and the laws applying to the state in which you live.

Based on research and information provided to The Wildlife Education and Research Foundation (WERF), I would make the following comments:

1. Live vaccines of any sort should not be used on *any* breed of canine.
2. Modified live and killed vaccines *do* work on Wolf Hybrids and many other species.
3. Whenever possible (especially with rabies vaccines) *killed* vaccines should be used to preclude the possibility of a virus "break" caused by the vaccine itself.

At Rudelhaus, we follow a strict and careful preventative medicine program and would not even consider skipping a booster when due— for their protection and for ours.

INTERNAL PARASITES

Worms, or infestations of internal parasites, can often be as deadly as the infectious diseases, particularly in puppies. If your puppy was wormed by the breeder, it is important to find out when it was wormed, because a second dose of the wormer must be administered approximately ten days later to take care of the eggs which were not hatched at the time of the first worming. If the puppy has been wormed and still shows symptoms of worms such as diarrhea, it should be verified by your veterinarian for proper treatment.

Dragging of the rectum on the ground may not be an indication of worms at all, but merely stimulation of the anal glands, a function formerly performed by the mother. The worm medicine should be prescribed by a veterinarian after determination, through microscopic examination of the stool (and if you are in an area where heartworm is present, blood sample) of the type of worms and the amount of medication needed. Take a fresh stool sample with you when you consult your veterinarian, as this is the only way he can determine the type of intestinal parasite present, or if there is one present at all.

SPECIAL WARNING ABOUT MEDICATIONS

For the most part, Wolf Hybrids can be diagnosed and treated for medical problems just like any other breed of dog. They are susceptible to the same diseases, parasites and stresses and the treatment for these problems is primarily the same.

There are some very essential differences between treatment of dogs and Hybrids, however, and both veterinarians and Hybrid owners should be aware of them.

SENSITIVITY TO MEDICATIONS

Wolf Hybrids are extremely sensitive to drugs in many cases. A high percentage Wolf Hybrid may be much more sensitive to various medications than another breed of dog of the same weight, and care should be taken that they do not receive overdoses of any medication. Many vets are tempted to give large doses of medication, tranquilizers, hormones and anaesthesia because of the size of the animal. In many cases, however, particularly with the higher percentage animals, a dosage which would seem perfectly reasonable for a dog of comparable size and weight, will send a Wolf Hybrid into shock. I have seen cases of animals being temporarily blinded by inadvertent overdoses of medication administered by unsuspecting vets who were not used to the sometimes super-senstivity to drugs and/or hormones. All of the high percentage animals I have had have been very nearly debilitated by a dosage of tranquilizer that would only calm a small Sheltie.

Conversely, one of the same animals was supposedly "knocked out" to have stitching done on its mouth, but when the critical moment arrived, "rose from a sound sleep" and sat up, staring the vet very threateningly in the eye. Later, after all the anaestheic had worn off, the lip was stitched with only a local anaesthetic and no problem at all.

A smaller dosage per weight of any medication seems to be indicated, although if the vet is experienced with your particular animal, the decision should be left to him or her. For those not accustomed to the animals' sensitivity, however, the word must be *caution*.

Despite the fact that wolves and Wolf Hybrids seem to eat and pass nearly anything, a radical change in diet, change of environment, altitude or temperature, or change of circumstances can produce diarrhea that can devastate the young pup and produce a prolonged case of diarrhea.

Worms seem to be more of a problem to a young Wolf Hybrid than other breeds and must be taken care of promptly and effectively. The diarrhea produced by the presence of the worms is much more difficult to control and eradicate than in domestic dogs.

Indeed, anything that upsets the digestive tract in the Wolf Hybrid produces a whole new set of symptoms that must be dealt with separately and often confuse the diagnosis of what is actually causing the problem for the pup.

When mild diarrhea or soft stools are a problem and worms or disease are not detected by the veterinarian, the animal should be put on a relatively bland diet—softened puppy chow (for the nutrients) and yogurt or cottage cheese and softened baby rice cereal (or the water that rice has been cooked in). Rice water and cheese products are long time standbys for firming stools in both animals and humans. Cottage cheese, besides being attractive in taste to an animal and a firming agent, alters the Ph factor in the intestines and thus helps in stabilizing the digestive tract.